BestMasters

Springer awards „BestMasters" to the best master's theses which have been completed at renowned universities in Germany, Austria, and Switzerland.

The studies received highest marks and were recommended for publication by supervisors. They address current issues from various fields of research in natural sciences, psychology, technology, and economics.

The series addresses practitioners as well as scientists and, in particular, offers guidance for early stage researchers.

Sven Herrmann

New Synthetic Routes to Polyoxometalate Containing Ionic Liquids

An Investigation of their Properties

Foreword by Prof. Dr. Carsten Streb

 **Springer** Spektrum

Sven Herrmann
Ulm, Germany

BestMasters
ISBN 978-3-658-08795-1   ISBN 978-3-658-08796-8 (eBook)
DOI 10.1007/978-3-658-08796-8

Library of Congress Control Number: 2015930457

Springer Spektrum

Printed on acid-free paper

Springer Spektrum is a brand of Springer Fachmedien Wiesbaden
Springer Fachmedien Wiesbaden is part of Springer Science+Business Media
(www.springer.com)

Foreword

In his Master project, Sven Herrmann carried out fundamental investigations into the development of polyoxometalate based ionic liquids (POM-ILs). For the first time he conducted a systematical study of POM-IL synthesis and properties, laying the foundations for a completely new field between inorganic supramolecular chemistry and nanomaterials chemistry. POM-ILs are obtained by charge balancing inorganic polyoxometalates (POM) anions with sterically demanding tetraalkylammonium or tetraalkylphosphonium cations. By functionalization of lacunary Keggin clusters with 3d-transition metals and charge balancing with tetraalkylammonium cations of varying chain length, a model system for the correlation of molecular structure with macroscopic materials properties such as viscosity and melting point was achieved. In this thesis, Sven Herrmann presents a systematic report on the syntheses via self-aggregation and ion-metathesis. Analytical methods for the characterization of the new compound class comprise UV-Vis, FTIR, NMR, EPR and Mößbauer spectroscopy. For the determination of important materials properties TGA and DSC were additionally carried out and rheological studies shed light onto the flow characteristics of the highly viscous materials. The thermal stability of the compounds was optimized by replacing ammonium with phosphonium-based cations. In preliminary studies Sven Herrmann investigated the reactivity of the materials towards azide and carbon monoxide, laying the basis to develop new materials for chemical and electrochemical small molecule activation.

During the course of his Master project, Sven Herrmann co-wrote a review article in the *Journal of Molecular and Engineering Materials* which summarizes the current state of the art in POM-IL chemistry and exemplifies potential fields of application for this novel class of compounds. In a recent publication in *Angewandte Chemie* "Polyoxometalate-based Ionic Liquids (POM-ILs) as Self-Repairing Acid-Resistant Corrosion Protection" (Angew. Chem. Int. Ed. 2014, DOI: 10.1002/anie.201408171) Sven reports the first example of POM-ILs as acid corrosion protection coating with self healing properties. The publication was highlighted by *Angewandte Chemie* as "hot paper" and is based on the results obtained during his Master project.

For his PhD project on POM-ILs Sven was awarded with a *Chemiefonds-Stipendium* fellowship by *Fonds der Chemischen Industrie*. His recent invention of a self-repairing acid corrosion protection was highlighted in the media (e.g. Südwest Presse SWP, Südwestdeutscher Rundfunk SWR, Innovations-Report.de, Chemie.de, Ingenieur.de, Chemical and Engineering News).

Prof. Carsten Streb

Institute Profile

Institute of Inorganic Chemistry I, Ulm University

The Institute of Inorganic Chemistry I at Ulm University is focused on the development of new catalysts and advanced functional materials for energy conversion and storage. Particular focus is the conversion of sunlight into chemical energy. The institute comprises the research groups of Prof. Carsten Streb (W3 o.L.) and Prof. Sven Rau (W3 m.L.)

The research group of Carsten Streb is focused on the development of new molecular metal oxides for use in energy conversion and energy storage, as molecular catalysts and photocatalysts, as well as materials applications on the interface of biological and inorganic chemistry.

Currently, one key aspect of the group's research is the use of solar energy to drive industrially important chemical oxidation reactions. Also, molecular catalysts are assembled which allow sustainable access to molecular hydrogen as energy carrier.

The group is also interested in supramolecular sensor systems which allow the detection of biologically relevant substrates in solution, preferably indicated by visual output signals.

Further, applications of molecular metal oxide-based ionic liquids in various material chemistry aspects such as corrosion protection or catalytic activity are currently under investigation.

The overarching research theme of the group is the development of new and general access routes which allow the predetermined synthesis and functionalization of molecular metal oxides, a challenging area where the group has recently made significant breakthroughs, giving access to new compound families with tuneable reactivity.

The group of Sven Rau develops molecular photosensitizers where sunlight is used to drive chemical redox processes. The group targets the development of intra- and intermolecular light-driven hydrogen evolution catalysts for sustainable energy systems.

A second research focus of the Rau group is the development of DNA-binding photoactive coordination compounds with applications in photodynamic therapy, e.g. for cancer treatment.

Both groups are linked by a deep interest in photochemical processes in complex chemical systems, and in particular the mechanistic study of interactions between ruthenium-based photosensitizers and molecular metal oxides in solution. Their joint research is focused on the detailed understanding of model systems with high relevance for the development of dye-sensitized solar cells and supramolecular photocatalysts.

Author Preamble

This work would not have been possible without the company of a big circle of friends going with me through the ups and downs a chemistry study implies. Above all would it not have been possible without the constant unquestioning support from my dear family. You are the prime pillar of my life and keep enriching it every day!

The present work was carried out between April and September 2013 at the Department of Chemistry and Pharmacy, Chair of General and Inorganic Chemistry of the Friedrich-Alexander-Universität Erlangen-Nürnberg headed by Prof. Karsten Meyer.

I wish to express my appreciation and thanks to **Prof. Carsten Streb** for the opportunity he offered me to carry out my Master project on a highly inspiring topic and to do so unforced, respecting my own research interests, with superb supervision, constant help and always his utmost kindness towards me - thank you Carsten!

Furthermore I would like to thank my laboratory colleagues for introducing me into laboratory routine concerning crucial synthetic and analytical procedures; namely **Andrey Seliverstov, Johannes Forster, Johannes Tucher** and **Katharina Kastner,** who had lots of helpful counsel when work had to be conducted and were great company in periods of free time. The same applies to **Stefanie Schönweiz** and **Benjamin Schwarz** my fellow campaigners.

Prof. Dr. Karsten Meyer, Chair of General and Inorganic Chemistry, I kindly acknowledge for giving me the opportunity to do research at his chair and providing his modern laboratory equipment including excellent analytical devices. His outstanding lectures were always highly instructive and fundamentally equipped me with the knowledge to understand the basic principles of inorganic chemistry. A good lecture is more than passing on knowledge - it is also about passing on passion. Prof. Dr. Karsten Meyer essentially contributed to strengthen my passion for the beauty of inorganic chemistry.

X

Furthermore my sincere gratitude goes to the fruitful cooperation with **Prof. Dr. Wierschem** and **Monika Kostrzewa** from the LSTM (Lehrstuhl für Strömungsmechanik) from the department of chemical and biological engineering of the FAU, for carrying out rheological investigations with my ionic liquids. The rheological data shown in this work were obtained by Monika.

A great thanks also goes to the many helping hands without whom a major research institute could not work. Be it staff carrying out routine measurements like **Christina Wronna** for elemental analyses, **Stephanie Bajus** and **Patrick Preuster** for thermogravimetric analyses or **Antigone Roth** for giving me access to various analytical equipment. Or be it **Manfred Weller** from the work shop or the glassblower **Ronny Wiefel** who could help me several times in realizing my experimental setups.

Henning Kropp I would like to thank for introducing me into the practical aspects of measuring EPR-spectra and **Dr. Jörg Sutter** for recording the Fe-57 Mössbauer spectra.

Dr. Matthias Moll (by now in his well earned retirement) and his entire staff I kindly acknowledge for their competent support in the handling of several analytical devices and their permanent readiness to help.

Sven Herrmann

Table of Contents

List of Abbreviations

POM	polyoxometalate
IL	ionic liquid
RTIL	room temperature ionic liquid
min	minute
h	hour
d	day
l	liter
ml	milliliter
M	molar
mmol	millimol
nm	nanometer
pm	picometer
Å	Ångström (= 10^{-10} m)
ε	extinction coefficient
mg	milligram
°C	degree Celsius
%	percent
e.g.	example given
m.p.	melting point
EA	elemental analysis
TGA	thermo gravimetric analysis
DSC	differential scanning calorimetry
w-%	weight percent
PBV	patent blue V
EPR	electron paramagnetic resonance
SQUID	superconducting quantum interference device
eq.	equivalents

Notation

In order to increase the clarity concerning extensive cluster formulas, the standard square bracket notation, e.g. $[SiW_{11}O_{39}Cu]^{6-}$ is simplified by an abbreviated notation using curly brackets, which enclose only the number and type of metal centers, e.g. $\{W_{11}Cu\}$, and represent the complete cluster unit.

Polyhedral subunits are indicated using the square bracket notation $[MO_x]$ (M = central transition metal, x = 4 − 6). In several occasions the assignment of the corresponding charge is missing due to the intention of giving only structural information concerning the polyhedral geometry.

Table of Compounds

entry No	formula	abbreviation
1	$(N(C_5H_{11})_4)_6[\alpha\text{-}SiW_{11}O_{39}Cu(H_2O)]$	$\{Q^5\}\{SiW_{11}Cu\}$
2	$(N(C_5H_{11})_4)_5[\alpha\text{-}SiW_{11}O_{39}Fe(H_2O)]$	$\{Q^5\}\{SiW_{11}Fe\}$
3	$(N(C_6H_{13})_4)_6[\alpha\text{-}SiW_{11}O_{39}Cu(H_2O)]$	$\{Q^6\}\{SiW_{11}Cu\}$
4	$(N(C_6H_{13})_4)_5[\alpha\text{-}SiW_{11}O_{39}Fe(H_2O)]$	$\{Q^6\}\{SiW_{11}Fe\}$
5	$(N(C_7H_{15})_4)_8[\alpha\text{-}SiW_{11}O_{39}]$	$\{Q^7\}\{SiW_{11}\}$
6	$(N(C_7H_{15})_4)_6[\alpha\text{-}SiW_{11}O_{39}Cu(H_2O)]$	$\{Q^7\}\{SiW_{11}Cu\}$
7	$(N(C_7H_{15})_4)_6[\alpha\text{-}SiW_{11}O_{39}Co(H_2O)]$	$\{Q^7\}\{SiW_{11}Co\}$
8	$(N(C_7H_{15})_4)_5[\alpha\text{-}SiW_{11}O_{39}Fe(H_2O)]$	$\{Q^7\}\{SiW_{11}Fe^{III}\}$
9	$(N(C_7H_{15})_4)_6[\alpha\text{-}SiW_{11}O_{39}Ni(H_2O)]$	$\{Q^7\}\{SiW_{11}Ni\}$
10	$(N(C_7H_{15})_4)_5[\alpha\text{-}SiW_{11}O_{39}Cr(H_2O)]$	$\{Q^7\}\{SiW_{11}Cr\}$
11	$(N(C_7H_{15})_4)_6[\alpha\text{-}SiW_{11}O_{39}Fe(H_2O)]$	$\{Q^7\}\{SiW_{11}Fe^{II}\}$
12	$(N(C_7H_{15})_4)_6[\alpha\text{-}SiW_{11}O_{39}Mn(H_2O)]$	$\{Q^7\}\{SiW_{11}Mn\}$
13	$(N(C_8H_{17})_4)_6[\alpha\text{-}SiW_{11}O_{39}Cu(H_2O)]$	$\{Q^8\}\{SiW_{11}Cu\}$
14	$(N(C_8H_{17})_4)_5[\alpha\text{-}SiW_{11}O_{39}Fe(H_2O)]$	$\{Q^8\}\{SiW_{11}Fe\}$
15	$K_6[\beta_2\text{-}P_2W_{18}O_{62}]19H_2O$	$\{\beta\text{-}P_2W_{18}\}$
16	$K_6[\alpha_2\text{-}P_2W_{18}O_{62}]19H_2O$	$\{\alpha\text{-}P_2W_{18}\}$
17	$(N(C_7H_{15})_4)_6[\alpha_2\text{-}P_2W_{18}O_{62}]$	$\{Q^7\}\{\alpha\text{-}P_2W_{18}\}$
18	$K_{10}[\alpha_2\text{-}P_2W_{17}O_{61}]20H_2O$	$\{\alpha\text{-}P_2W_{17}\}$
19	$(N(C_7H_{15})_4)_8[\alpha_2\text{-}P_2W_{17}O_{61}Co(H_2O)]$	$\{Q^7\}\{\alpha\text{-}P_2W_{17}Co\}$
20	$(^{6,6,6,14}P)_6[\alpha\text{-}SiW_{11}O_{39}Cu(H_2O)]$	$(^{6,6,6,14}P)\{SiW_{11}Cu\}$
21	$(^{6,6,6,14}P)_5[\alpha\text{-}SiW_{11}O_{39}Fe(H_2O)]$	$(^{6,6,6,14}P)\{SiW_{11}Fe\}$
22	$(N(C_7H_{15})_4)_7[\alpha\text{-}BW_{11}O_{39}Cu(H_2O)]$	$\{Q^7\}\{BW_{11}Cu\}$
23	$(N(C_7H_{15})_4)_6[\alpha\text{-}BW_{11}O_{39}Fe(H_2O)]$	$\{Q^7\}\{BW_{11}Fe\}$
24	$(N(C_7H_{15})_4)_7[\alpha\text{-}PW_{11}O_{39}Cu(H_2O)]$	$\{Q^7\}\{PW_{11}Cu\}$
25	$(N(C_7H_{15})_4)_6[\alpha\text{-}PW_{11}O_{39}Fe(H_2O)]$	$\{Q^7\}\{PW_{11}Fe\}$
26	$K_8[\gamma\text{-}SiW_{10}O_{36}]$	$\{SiW_{10}\}$
27	$(N(C_7H_{15})_4)_4[\gamma\text{-}SiW_{10}O_{36}]$	$\{Q^7\}\{SiW_{10}Cu_2\}$

Publications

The following articles and communications were published as a result of work undertaken during the years of study or in the course of this master thesis:

(1)   "On the mechanisms of ionic conductivity in $BaLiF_3$: a molecular dynamics study", D. Zahn, S. Herrmann, P. Heitjans, *Phys. Chem. Chem. Phys.*, **2011**, 13, 21492–21495

(2)   "Polyoxometalate ionic liquids (POM-ILs) - The ultimate soft polyoxometalates? A critical perspective", S. Herrmann, A. Seliverstov, C. Streb, *Journal of Molecular and Engineering Materials*, **2014**, 2,1440001-1440007

(3)   "Polyoxometalate-based Ionic Liquids (POM-ILs) as Self-Repairing Acid-Resistant Corrosion Protection" S. Herrmann, M. Kostrzewa, A. Wierschem, C. Streb, *Angew. Chem. Int. Ed,.* **2014**, DOI: 10.1002/anie.201408171

1. Introduction
1.1 Ionic liquids - past, present and the way to go

Ionic liquids (ILs) are a remarkable compound class, facing back on a short yet prospering history as unique solvents and electrolyte materials. In recent years there has been an actual explosion of interest in them and ILs sparked a tremendous amount of scientific research. Leading scientific companies coined phrases like "Ionic Liquids - Solutions for Your Success" (*BASF basionics*), "Ionic Liquids - Don´t Miss Your Opportunity to Innovate" (*Merck COIL 1, 2005*) or "Ionic Liquids - Enabling Technologies" (*Sigma-Aldrich (ChemFiles Vol. 5(6))*. But before starting to take a closer look on the reasons for the triumphal course of ILs a short retrospect on their history shall be given. Pioneering work in the field of ionic liquids dates back to 1800 when Humphrey Davy used the then-newly invented voltaic pile to pass electricity through molten salts of alkali metals in order to obtain sodium and potassium, later also magnesium, calcium, barium and strontium under the influence of the electric field.[1]

Figure 1 Latvian stamp honors anniversary of Walden's birth.

In 1914 Paul Walden described the synthesis and properties of the "first" ionic liquid, ethylammonium nitrate, featuring a melting point of 12 °C. According to his definition, ionic liquids are "materials composed of cations and anions, that melt around 100 °C or below as an arbitrary temperature limit", which sets them apart from molten salts that exhibit higher melting points.[2, 3] The initial research of ionic liquids was mainly centered around organic compounds with asymmetric alkyl chains that effectively prevented formation of lattice structures and therefore ensured that those compounds were liquid even below 100 °C.

2

Ionic liquids of the first generation were yet prone to hydrolyze as illustrated by the frequently used $[AlCl_4]^-$ anion. Thus ionic liquids did have a barely regarded shadowy existence as laboratory oddity for a long time - and this despite their appealing properties. With the rediscovery of stable anions and the development of the second generation of ionic liquids in the aftermath the interest in ionic liquids started to grow constantly. Particularly during the past 20 years ionic liquids received more and more interest due to special characteristics like their low melting point and their extremely low vapor pressure as well as their tunable properties making them suitable for a broad field of application. For instance in batteries or fuel cells, when conventional solvents fail, the use of ionic liquids can deliver convincing results. Ionic liquids can even be applied in biology and biomimetic processes to improve enzymatic activity.[4-7] Over the last decade research concerning ionic liquids experienced a boom with an exponential increase in publications and patents (see Figure 2).

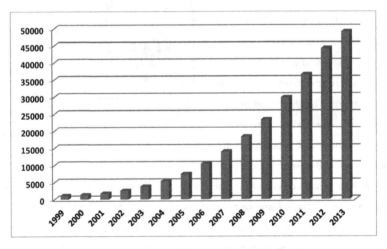

Figure 2 Number of total publications in the last 15 years containing "ionic liquid" as concept. Data taken from the SciFinder search function.

A classical example of an IL is 1-ethyl-3-methyl-imidazolium ethylsulfate (m.p. < - 20 °C) which has a significantly lower symmetry than a classic inorganic salt e.g. like table salt NaCl (m.p. 801 °C). Consequently, solidification of the ionic liquid will take place at lower temperatures. The strong ionic interaction within ILs results in a negligible vapor pressure, unless decomposition occurs. It makes the material non-flammable and highly stable thermally, mechanically and electrochemically. Furthermore, it can impart very appealing solvent properties and immiscibility with water or organic solvents which can be used to create biphasic systems. Some key-properties of ionic liquids are summarized in Table 1.[8, 9]

Property	typical values
Melting point	< 100 °C
Liquidus range	> 100-200 °C
Viscosity	< 100 - 150 cP
Thermal stability	High
Vapor pressure	Negligible
Dielectric constant	< 20 - 40
Specific conductivity	< 10 - 15 mS cm^{-1}
Electrochemical window	Up to 6 V

Table 1 Key properties of modern organocation-based ionic liquids.[8, 9]

A general synthetic approach for obtaining ionic liquids is the pairing of sterically demanding cations with a range of organic or inorganic anions. Most modern ionic liquids are formed when bulky organic cations such as alkylammonium, alkylpyridinium or imidazolium cations are combined with a range of weakly interacting organic or inorganic anions such as Cl$^-$, BF$_4^-$ or PF$_6^-$ or CF$_3$SO$_3^{2-}$. Due to their unrivalled properties (high thermal/chemical stability in combination with low vapour pressure and good solvent properties) ILs have become one of the fastest growing fields of research in recent years. The kind of ions used in a ionic liquid essentially influence its viscosity, solubility behavior, melting point as well as its thermal and electrical stability. For this reason ILs made a name for themselves as "designer solvents".[10] Some of the typical ions used are shown in Figure 3 on the next page.

4

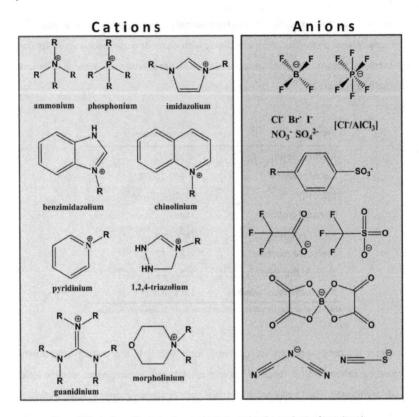

Figure 3 Illustration of typical cations and anions used in the synthesis of ionic liquids.

An important sub-class of ionic liquids are room-temperature ionic liquids (RTILs) [8, 9] which combine the usefulness of classical organic solvents with the benefits of ionic liquids. Since the discovery of imidazolium-based RTILs, a wide range of technical applications has been developed, most notably in chemical synthesis,[11] electrochemistry,[12-15] catalysis [8, 9, 16-19] and sensor systems [20-22].

1.2 Applying clusters in ionic liquids - the creation of composite materials

The deliberate synthesis of multifunctional materials from well-defined building blocks of a (virtual) library is one of the most challenging problems in contemporary chemistry. Combining the properties of organic and inorganic components in one compound offers access to such materials, often featuring unique physical and chemical properties. The concept itself is quite old and widely used in industry where functional inorganic pigments are dispersed in organic media or when rubber is stabilized by addition of carbon black in the manufacturing of tires. In current research, this approach is used in the combination of bulky organic cations with (often) inorganic anions to give composite ionic liquids. Typically, IL property tuning is achieved by modification of the organic cation, whereas little is known about anionic functionalization. Recently, this challenge has been addressed when anionic molecular metal oxides, so-called polyoxometalates (POMs) were employed as versatile anionic components, giving rise to polyoxometalate-based ionic liquids (POM-ILs) and thus advancing the frontiers of materials science. Figure 4 (see below) shows a schematic picture of a POM-IL on the molecular scale with Keggin POMs as inorganic component charge balanced with tetrahedral organo-cations.

Figure 4 Polyoxometalate Keggin clusters (blue polyhedra) paired with tetrahedral organic counter-ions like tetraalkylammonium or tetraalkylphosphonium ions (purple tetrahedra) to yield the POM-IL as an organic-inorganic composite material.

Before starting to shed some light onto the merits POM-IL chemistry brought forward, a short introduction into polyoxometalates in general and their chemistry in particular is given in the following chapter.

1.3 Properties and formation of polyoxometalates (POMs)
1.3.1 General aspects and history of POMs

Polyoxometalates (POMs) are inorganic metal-oxo cluster anions of the general formula $[M_xO_y]^{n-}$ consisting of two or more high-valent transition metals (M) (e.g. M = W, V, Mo, Nb, Ta) which are linked together via oxo-ligands (O). The early transition metals are incorporated into the cluster as oxoanions in a high oxidation state (d^0, d^1) and are called the addenda atoms, which together with the oxo-ligands form coordination polyhedra of the type $[MO_y]$ (y = 4-7). The formation of polyoxometalates follows a unique self-assembly process of these preformed building blocks which can be linked by sharing corners, edges and more rarely faces (see Figure 5). The two most common linking modes are the corner sharing mode with one bridging μ-oxo-ligand and the edge-sharing mode with two bridging μ-oxo-ligands.

Figure 5 Illustration of the three common coordination modes of $[MO_6]$ octahedral units. Left: corner-sharing mode. Middle: coordination via edge-sharing mode. Right: face-sharing of two $[MO_6]$ octahedra. Color scheme: *addenda* atoms M = e. g. Mo^{VI}, W^{VI}: blue, O red.

The first polyoxometalate reported was a phosphomolybdate, namely $[PMo_{12}O_{40}]^{3-}$, which was obtained by *Berzelius* in 1826 from ammonium molybdate $(NH_4)_2MoO_4$ with an excess of phosphoric acid.[23] Nearly forty years later in 1864, *Marignac* synthesized the isostructural silicotungstic acid $H_4[SiW_{12}O_{40}]$. He observed two different forms of it, which later turned out to be two structural isomers - today referred to as the α– and β– form.[24] Solving the structure of the new compound class required more sophisticated analytic methods than were available in the 19th century. Only with X-ray diffraction, discovered by *Max von Laue* in 1912 [25] and pioneered by *W. H. Bragg* and his son *W. L. Bragg* [26], it was possible to gain clarity of the structure. It was the chemist *J. F. Keggin* who experimentally solved the α-isomer structure in 1933 and thus lent his name to this class of cluster-compounds: the Keggin-clusters.[27] Nowadays a broad structural diversity of clusters is known beside the Keggin cluster (see. Figure 6).

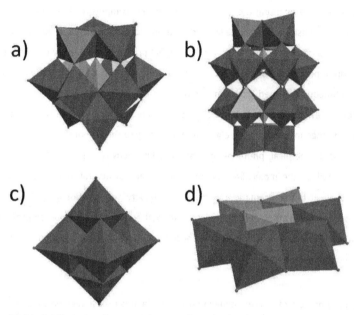

Figure 6 Polyhedral illustrations for some polyoxometalates showing their structural diversity. a) Keggin structur $[XM_{12}O_{40}]^{n-}$; b) Dawson structure $[X_2M_{18}O_{62}]^{n-}$; c) Lindqvist hexamolybdate $[Mo_6O_{19}]^{2-}$; d) Anderson structure $[XM_6O_{24}]^{n-}$. Color scheme: addenda atoms M = e. g. Mo^{VI}, W^{VI}, V^V: blue, templating metal X = e.g. B, Si, P: orange, O red.

POMs may also include a variety of heteroatoms ranging from main-group metals (e.g. X = P, As, Si, Ge) to d-elements (e.g. Cu, Fe, Ni, Mn, Cr) and even lanthanides (e.g. Ce, La, Sm), turning POMs into one of the structural most diverse inorganic compound class. They exhibit an almost unmatched range of physical and chemical properties with a rich chemistry, structural diversity and the ability to form systems ranging in size from the nano- to the micrometer scale.[28] For this reason POMs have attracted growing interest to the scientific community during the last few decades with relevance to analytical chemistry [29], catalysis [30, 31] and medicine [32-34] and are as novel materials of great interest to materials science.

1.3.2 Formation of clusters - the self-assembly process

Mimicking "natura naturans" or the self-causing activity of nature is a long cherished dream of chemists. In nature complex molecular systems like proteins are formed by molecular growth guarding the gateway to molecular complexity. Attaining extensive compounds correspondingly in the lab under one-pot conditions without the problem of repetitive separation and purification of individual products is a fiercely pursued goal. The world of polyoxometalates is such a kind of chemistry were highly ordered clusters are spontaneously formed from smaller building blocks by self-assembly, leading to completely new physical and chemical properties compared to the starting materials. POMs can range in size from merely six metal centers to the size of many proteins and show versatile chemical, photochemical and catalytic activity. Their properties can be further tuned on the molecular level by chemical and structural modification of the cluster shell. It is this dramatic increase in complexity and functionality without the need of extensive synthetic efforts which makes self-assembled structures so promising for materials science and to one of the most fascinating areas of research in modern inorganic chemistry.

Though the process of self-assembly in polyoxometalates is described very well,[35] and it has been studied extensively by in-situ ^{1}H-, ^{17}O- and ^{138}W-nuclear magnetic resonance spectroscopy, the exact reaction mechanism is still unknown.[36-39] However it is clear that the assembly of large cluster compounds, falling into the field of inorganic coordination chemistry, is strongly influenced by coordinative bonds owing to their kinetic lability and variable bond strength. In particular the self-assembly of polyoxometalates, anionic transition metal-oxygen clusters, is dominated by the formation of coordinative bonds between the electron deficient metal centre and oxygen atoms which act as nucleophilic ligands. Assembly of these preformed complexes into structurally extended systems is driven mainly by intra- and intermolecular interactions such as Coulomb interactions, hydrogen bonds and dispersive interactions which guide the self-assembly into complex systems and which enable the formation of a multitude of different supramolecular structures.[40] One main goal for synthetic chemists is a deeper understanding of this self-organizing process to ideally direct it to the incorporation of certain well known structural elements.

POM formation is best illustrated using the archetypal Keggin anion $[XM_{12}O_{40}]^{n-}$ (X = B, Si, P, etc.; M = W, Mo). POMs in general, and the Keggin anion in particular are prepared by acidifying an aqueous solution of an oxometalate precursor such as Na_2MO_4 (M = W, Mo). The decrease of pH results in an expansion of the metal coordination shell of the starting materials from a tetrahedral to an octahedral coordination sphere leading to an increased polarization of the M=O double bonds. Afterwards the terminal oxygen atoms can be protonated more easily, initiating the onset of consecutive condensation reactions between the $[MO_6]$ units which are hereby linked by one or two bridging μ-oxo-ligands respectively and react in corner- and edge-sharing modes (see Scheme 1). Larger polynuclear building blocks are formed from highly reactive hydrated complexes such as $O[MO(OH)_4]_2^{3-}$.[41]

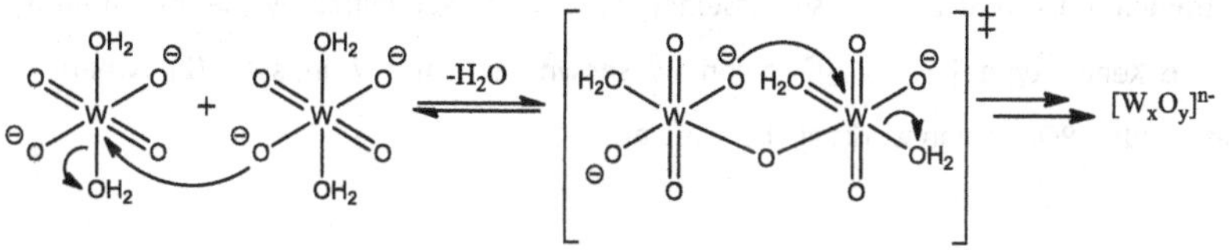

Scheme 1 Condensation reaction of $[WO_6]$-octahedra to larger polynuclear $[W_xO_y]^{n-}$ building blocks.

These unstable intermediates can immediately react further and aggregate to highly symmetric ($C3_v$) edge-sharing triads $[W_3O_{13}]$. The final framework is obtained by further fusion of the unstable hydrated polynuclear metal-oxide intermediates. Hereby the cluster growth is essentially influenced by empty d-orbitals of the high oxidation state metals that enable the formation of strong π − bonds, thus decreasing basicity and nucleophilicity of the terminal M=O double bonds. Consequently further spontaneous condensation reactions are prevented and the cluster growth comes to an end - the final cluster structure is completed.

Reasons for this phenomenon and the enormous variety of compounds that can be formed by linking together metal-oxide building blocks are the easy change of coordination numbers as well as easy exchange of water ligands at the addenda atoms, the moderate strength of Mo-O-Mo type bonds allowing "split and link" type processes, the easy exchange of electron densities without a strong tendency to form metal-metal bonds and the presence of the terminal M=O groups preventing in principle unlimited growth to extended structures.[42]

1.3.3 Template assisted cluster formation

One very well reviewed structure exemplifying a self-assembly condensation process is that of the Keggin polyoxometalate type $[XM_{12}O_{40}]^{n-}$ (M = Mo, W; X = B^{3+}, Si^{4+}, P^{5+}). This cluster-type will be used in the following to more precisely illustrate the key principles of a templated self-assembly process of POM-frameworks.

In case of the Keggin structure, oxoanions such as borate, silicate or phosphate can act as templates and allow the controlled Keggin anion synthesis. The typical Keggin structure with T_d symmetry is then formed of four aggregating triads templated by a tetrahedral anion XO_4^{n-}. This self-assembly process is exemplified by the formation of the α-Keggin-type $[SiW_{12}O_{40}]^{8-}$ anion as shown in Figure 7 in step (2), where a tetrahedral PO_4^{3-} anion works as the template.

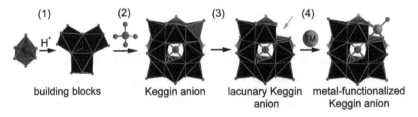

(1)	(2)	(3)	(4)
building blocks	Keggin anion	lacunary Keggin anion	metal-functionalized Keggin anion

Figure 7 Self-assembly of the Keggin anion $[\alpha\text{-}SiW_{12}O_{40}]^{4-}$. (1) Oxometalate building blocks (e.g. $\{WO_6\}$, $\{W_3O_{13}\}$) are spontaneously formed in solution by acidification. (2) In the presence of tetrahedral templates (e.g. SiO_4^{4-}), the building blocks assemble to give the Keggin anion $[\alpha\text{-}SiW_{12}O_{40}]^{4-}$. (3) Removal of one W center leads to the formation of a lacunary species, $[SiW_{11}O_{39}]^{8-}$ (the structural vacancy is marked with an orange arrow). (4) Transition metal (M) incorporation into the structural vacancy leads to the formation of a TM-functionalized Keggin anion $[SiW_{11}O_{39}M(H_2O)]^{n-}$. Color scheme: W (black), O (red), Si (purple).

The Keggin architecture can be further modified by controlled hydrolysis which allows the selective removal of one or several centers, giving access to so-called lacunary species such as $[SiW_{11}O_{39}]^{n-}$, see Figure 7 step (3). Lacunary clusters feature one or several metal binding sites. Incorporation of functional metals into these binding sites can be used to tune the properties of the Keggin anion, giving species of the type $[SiW_{11}O_{39}M(H_2O)]^{n-}$, see Figure 7 step (4). Further modification of the cluster is possible by coordination of various ligands beside water to the transition metal in a ligand substitution reaction.[43] An overview of the structural diversity arising out of the manifold motives of Keggin-based clusters is shown below in Figure 8

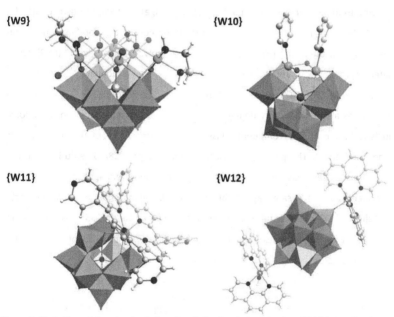

{W9}   {W10}

{W11}   {W12}

Figure 8 Illustration of the structural diversity of the Keggin-based anions highlighting the diverse coordination environments which are determined by the number of vacancies and by the type of isomer. Top left: the trilacunary vacancy of $[SiW_9O_{34}]$ is filled up by a $[Ni_6O_9(en)_3]$ framework ; top right: $[SiW_{10}O_{38}Al_2]^{6-}$ with one pyridine ligand bound at each Al^{3+} centre; bottom left: $[PW_{11}O_{39}]^{7-}$ bound over its lacunary site to hafnium chelated by a porphyrin-system; bottom right: $[PW_{12}O_{40}]^{3-}$ bridging two cobaltphenanthroline complexes. Color scheme: cluster: light blue polyhedra, O: red, Si: light grey, P: pink, N: dark blue, C: grey, Al: lime, Ni: turquoise, Co: green, Hf: orange.

1.4 Merging ILs with POMs - the creation of POM-ILs

1.4.1 General aspects and background of POM-ILs

After the separate introduction into ILs and POMs respectively this chapter exemplifies advantages as well as limitations of merging both fields into composite materials, where polyoxometalate clusters act as the anionic component of an ionic liquid, leading to POM-ILs. The POM clusters can be combined with a range of bulky organic cations to give versatile materials which are liquid at or around room temperature - so called room temperature ionic liquids (RTILs).

Over the last decade, several groups have addressed the purposeful synthesis of POM-ILs. Some examples are given in Figure 9 (see next page). In 2004, the first report of a true POM-IL was published.[44] The material is based on the combination of the archetypal Keggin-anion $[PW_{12}O_{40}]^{3-}$ with the polyethylene glycol (PEG)-based quaternary ammonium cation Ethoquad 18/25, $[(CH_3)(C_{18}H_{37})N[(CH_2CH_2O)_nH][(CH_2CH_2O)_mH]^+$ (= Q+) (See **POM-IL-1** in Figure 9). The Ethoquad as a sterically highly demanding cation predominantly determines the ionic liquids behavior of this compound class, whereas the choice of anion has a less pronounced effect. The parent cation compound Ethoquad 18/25 featuring chloride counterions is itself a viscous material at room temperature and applying it as cationic component is a relative safe way for obtaining ionic liquid materials as previous reports could show.[14] POM-IL-1 features a glass transition temperature of -35 °C, is stable upon heating up to ca. 160 °C and shows an unusually high ionic conductivity of Λ ca. 6×10^{-4} S cm^{-1} at 140 °C.

True POM-ILs need to be distinguished from a simple polyoxometalate cluster salt which has been dissolved in an ionic liquid. Firstly, the dissolution of a POM salt in ILs is limited by solubility and therefore true POM-ILs can achieve much higher cluster loadings (as the POM cluster acts as the anionic component of the IL). Further, dissolution of POM salts in ILs always results in the introduction of two additional ionic species (i.e. the cationic part of the POM salt and the anionic part of the IL) which will change the physical and chemical behavior of the mixture, thus leading to inadvertent effects, particularly at higher POM loadings. For this reason the synthesis of true POM-ILs is so auspicious.

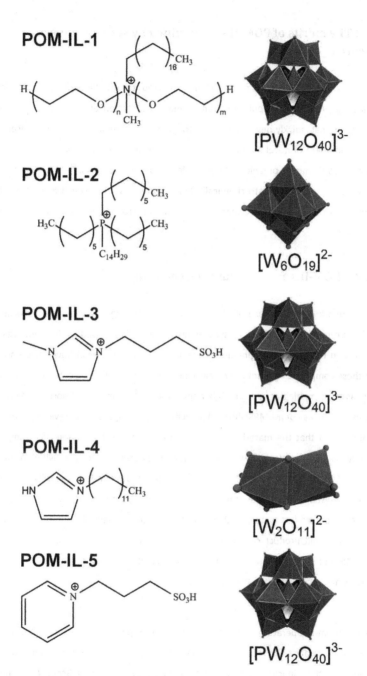

POM-IL-1

$[PW_{12}O_{40}]^{3-}$

POM-IL-2

$[W_6O_{19}]^{2-}$

POM-IL-3

$[PW_{12}O_{40}]^{3-}$

POM-IL-4

$[W_2O_{11}]^{2-}$

POM-IL-5

$[PW_{12}O_{40}]^{3-}$

Figure 9 Examples in literature for POM-ILs. For detailed information see indicated reference.

1.4.2 The merits of POM-ILs exemplified by state-of-the-art research

Until recently, POMs have rarely been connected to the field of soft materials. Pioneering work on soft polyoxometalate materials started at the end of the 20th century and has quickly developed into a thriving field. Research has been focused on the development of novel materials which synergistically combine "hard" POMs with a range of "soft", mostly organic materials to form novel compounds.[45-47] Their usage as catalysts, sensors and electrochemically active materials is shortly discussed in this chapter, with three literature examples to give insight into the current frontiers of POM-IL research.

1.4.2.1 POM-ILs as self separating catalysts

In a landmark paper, the use of POM-ILs as self-separating catalysts for esterification has been reported.[48] The authors synthesized three POM-ILs based on different organic cations and the plenary Keggin- cluster anion $[PW_{12}O_{40}]^{3-}$. It should however be noted that these compounds are not truly ionic liquids but should rather be considered molten salts, as their melting points are higher than 100 °C.[48-50] In their report, the authors focused on the use of $[MIMPS]_3[PW_{12}O_{40}]$ (POM-IL-3; see Figure 9 on previous page) and demonstrated that the material effectively catalyzes the esterification of a range of carboxylic acids and alcohols with almost quantitative yields and selectivities of more than 90 %. Further, the material can easily be separated: **POM-IL-3** is readily dissolved in the polar starting materials (carboxylic acid, alcohol) but is immiscible with the less polar reaction product (ester) so that after the reaction, a biphasic system is obtained where the liquid product can easily be separated from the solid catalyst. Thus, the advantages of homogeneous and heterogeneous catalysis can be combined, facilitating recovery and recycling of the POM-IL catalyst.

A true room-temperature POM-IL was recently reported as an excellent liquid epoxidation catalyst. The compound, **POM-IL-4**, is based on n-dodecylimidazolium cations and the dinuclear peroxotungstate cluster $[W_2O_{11}]^{2-}$.[51] **POM-IL-4** has been shown to be active for the epoxidation of olefins and allylic alcohols with almost quantitative yield and high selectivity. Further studies compared **POM-IL-4** with three related POM-ILs and a correlation between the chemical nature of the organic cation and the catalytic performance of the catalyst was found, highlighting the importance of both constituents in POM-IL-based catalytic systems.[51] An efficient reaction-induced phase-separation system could be developed by the authors in which the reaction system can switch from tri-phase to emulsion and then to biphase and finally to all the catalyst self-precipitating at the end of the reaction. Again recovery and reuse of the present catalyst is facilitated.

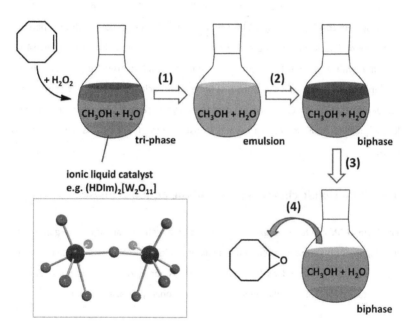

Figure 10 Illustration of POM-ILs applied for improved catalyst recovery, demonstrated at hand of the "reaction-induced self-separating catalyst" POM-IL-4, which is applied for the epoxidation of cyclooctene with H_2O_2. Before the reaction, there are three separate phases in the flask: the IL-catalyst as bottom layer (dark green), the solvent MeOH/water (1:3 volume ratio) (light green) and the organic cyclooctene layer (purple). At the beginning of the reaction emulsification of the system takes place (step 1). After a short time two phases form and the catalyst is all dissolved in the substrate, while the mixed solvent is almost transparent (step 2). At the end of the reaction all the catalyst self-precipitates at the bottom, when the H_2O_2 is used up (step 3). Separation of the catalyst from the product is simply achieved by decantation (step 4). The catalyst anion $[W_2O_{11}]^{2-}$ is depicted in the box on the lower left. Color scheme: tungsten (blue), oxygen (red).

1.4.2.2 POM-ILs applied for industrial processes

The application of **POM-IL-3** was even further extended and a recent study has highlighted the use of this POM-IL for the hydrothermal catalytic conversion of cellulose into a range of useful fine chemicals, thereby demonstrating the assets of POM-ILs in synthetic industrial processes.[52]

Another example is the deep desulfurization of model diesel fuels which has been demonstrated using a Keggin-based POM-IL featuring the *N*-propanesulfone pyridinium (PsPy$^+$) cation.[53] The compound, **POM-IL-5**, (PsPy)$_3$[PW$_{12}$O$_{40}$] was employed in a three-phase system consisting of the POM-IL as bottom layer, an aqueous hydrogen peroxide phase as middle layer and the diesel oil as top layer. It was shown using **POM-IL-5** together with a standard IL, [omim]PF$_6$, almost complete oxidative removal of the model sulphur pollutant dibenzothiophene was achieved within 1 hour at a temperature of T = 30 °C. The system could be recycled several times with only marginal loss of activity, thus demonstrating the industrial importance of this new compound class.

1.4.2.3 Inorganic chemistry beyond the aqueous phase

Employing POMs as fully inorganic compounds with high crystal lattice energies in ionic liquids, drastically expands their applicability as it allows dissolution in organic solvents. POMs may thus be used as homogenous catalyst or Brønsted acids with reactions necessitating organic media which is not or only partially possible with many "traditional" POMs.

Beside increasing their field of application the ionic liquid medium can exhibit altered chemical conditions and therefore change the reactivity of POMs. A remarkable example is the photochemical oxidation of water carried out at interfaces of water with ionic liquids in the working group of *Alan M. Bond*.[54] [P$_2$W$_{18}$O$_{62}$]$^{6-}$ is photochemically reduced in "wet" ionic liquids like 1-butyl-3-methylimidazolium tetrafluoroborate [BMIM][BF$_4$] where water acts as an electron donor and is itself oxidized to dioxygen. Such type of redoxreaction does not occur if the [P$_2$W$_{18}$O$_{62}$]$^{6-}$ is dissolved in neat water or in wet molecular organic solvents. The structure of the interfacial water is expected to differ from bulk water which most likely leads to this unexpected reactivity.[54] Ionic liquid media in general have dramatic effects on the redox-potentials of substrates why further intriguing discoveries may be expected from this young field of research.[54, 55]

2. Objective

Motivated by promising examples of improvements brought forward by the use of ionic liquids in current state of the art research (see chapter 1.4.2) concerning extended solvation properties, improved recyclability of catalysts or altered chemical behavior, was the main focus of this work directed to implement existing cluster structures into organic media by conversion into ionic liquids. The long term goal is to create unprecedented materials with unique properties such as high viscosity, excellent conductivity and thermal stability - interesting for both material science and synthetic processes with potential usage as catalysts, sensor systems or electrochemically active materials.

As nearly any element can be embedded in a polyoxometalate their compound family is already one of the most versatile available. However polyoxometalates as inorganic salts with high crystal lattice energies are faced with low solubility in non-polar organic solvents restricting their application for catalytic reactions in those media. Conversion of crystalline POMs into ionic liquids may overcome this limitation and may just bring polyoxometalate chemistry to its full potential.

POMs dissolved in ionic liquids as solvents can therefore increasingly be found in literature. [56-58] Such a straight forward approach may suffice for many applications yet be insufficient for others as dissolution of POM salts in ILs always results in the introduction of two additional ionic species (i.e. the cationic part of the POM salt and the anionic part of the IL) which will alter the physical and chemical behavior of the mixture, thus leading to inadvertent effects, particularly at higher POM loadings (see Figure 11 on the next page).

POM-IL   **POM dissolved in IL**

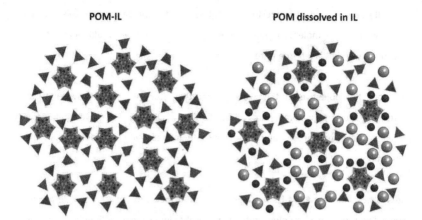

Figure 11 Molecular model for a true POM-IL (*left*) vs. a POM dissolved in IL (*right*). The POM-IL exhibits a much higher cluster density and has less different electrostatic interactions. In a POM-IL only interactions among POMs (blue polyhedra with templating heteroelement in pink) and their charge balancing cations (purple tetrahedrals) are possible, whereas for POMs dissolved in a ionic liquid additional interactions with the ionic liquids cations (black balls) and anions (grey balls) occur.

electrostatic interactions   **electrostatic interactions**
POM-IL   **POM dissolved in IL**

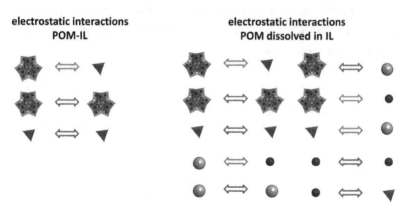

Figure 12 All possible types of electrostatic interactions among the ions of a true POM-IL vs. those of a POM-salt dissolved in IL. Attractive interactions are indicated with a green double arrow, repulsive forces with a red one.

Even if no negative effects arise due to the additional solvent ions and their manifold interaction possibilities (see. Figure 12) is the catalytic activity restricted because dissolution of a POM salt in ILs is limited by solubility. In contrast, true POM-ILs can achieve much higher cluster loadings (as the POM cluster acts as the anionic component of the IL) being the ideal supported reagent system for the catalyst (in this case the POM) where each molecular unit of the supporting material would bear inherent functionality and be itself free to react.

Therefore the objective of this work and further studies is set to systematically synthesize true polyoxometalate ionic liquids (POM-ILs), to investigate their physiochemical properties and to probe their reactivity fathoming potential applications.

The synthesis of new POM-ILs is supposed to be achieved using sterically demanding cations like tetraalkylammonium, tetraalkylphosphonium or imidazolium ions featuring long alkyl chains. Tunability of the ILs properties is further realized by alternation of the heteroatoms and incorporation of different d-elements into the lacunary site(s) of the mono- or polyvacant cluster anions. Basic characterization of the obtained products is carried out by elemental analysis (EA), infrared spectroscopy (IR) and UV-Vis spectroscopy. Rheological measurements are carried out in cooperation with the LSTM (Lehrstuhl für Strömungsmechanik) to complement the investigation of physical properties of the obtained new ILs.

3. Results and Discussion
3.1. Ionic liquids based on Keggin clusters

In this work a systematic investigation of how the alkyl chain length of the cations effects the resulting compounds properties was conducted. As model system worked the archetypal Keggin-cluster as monolacunary anion with Cu^{II} and Fe^{III} embedded as additional transition metal in the lacuna. This cluster family is an ideal prototype as it offers wide coordination flexibility,[59] good solution stability [56] and facile synthetic access.[56] The cluster was charge balanced with tetraalkylammonium cations Q^x (x denotes alkyl chain length, from $n\text{-}(C_5H_{11})_4N^+$ (= Q^5) to $n\text{-}(C_8H_{19})_4N^+$ (= Q^8) (1,2,3,4,6,8,13,14).

As the cluster charge determines the number of counter cations does it have important influence on many physical parameters (viscosity, melting point, etc.). A convenient way of influencing the charge of the cluster anion without applying completely different cluster types is to vary the templating heteroelement or the transition metal incorporated in the lacunary site, both of which are also of interest with respect to the chemical reactivity. In this work six transition metal (TM)-substituted Keggin anions of the type $[SiW_{11}O_{39}TM(H_2O)]^{n-}$ (TM =Cr^{III}, Fe^{III}, Co^{II}, Ni^{II}, Cu^{II}, Mn^{II}) were synthesized and investigated for this purpose (6-11). For heteroelement variation Keggin anions of the type $[XW_{11}O_{39}TM(H_2O)]^{n-}$ ($X = B^{3+}$, Si^{4+}, P^+) were applied, substituted with Cu^{II} (22, 6, 24) and Fe^{III} (23, 8, 25) respectively. Tetraheptylammonium ions (= Q^7) were applied as counterions for this investigation.

3.1.1 Structure and synthesis of the Si-templated Keggin clusters

Early transition metal polyoxoanions have traditionally posed serious challenges to the synthetic chemist because of their stoichiometric and structural complexity. However over the past decades an increasingly systematic approach to their synthesis was developed mainly based on empiric observations.[56]
A compound class of clusters that is well investigated with well established synthetic routs is those of Keggin clusters based on tungsten or molybdenum as the addenda

atom. Simple metal oxide precursors such as molybdenum(VI) oxide, MoO_3 or sodium molybdate, Na_2MoO_4 are dissolved in aqueous media. Acidification of the solution leads to an expansion of the coordination shell from 4 to 6, from a tetrahedral to an octahedral oxygen coordination environment. This in turn enables protonation of the terminal oxygens and facilitates condensation reactions between the particular $[MO_6]$ fragments. As a result, larger structures are self-assembled in solution.

As precursor for the ionic liquids based on Keggin-type clusters the $K_8[\alpha\text{-}SiW_{11}O_{39}]\cdot13H_2O$ was used. It is derived from the Keggin anion $[SiW_{12}O_{40}]^{4-}$ (see Figure 13 left part), which structurally belongs to a group of clusters that are made up by W_nO_{3n} units encapsulating one or more formally anionic subunits. The main network of this cluster is built of a $W_{12}O_{36}$ cage and therefore of four independent $[W_3O_9]$ subunits plus $[SiO_4]^{4-}$ encapsulated as tetrahedral oxoanion. This leads effectively to a truncated tetrahedron of T_d-symmetry (see Figure 13 right part with the representation boiled down to the central tetrahedron drawn into the dodecanuclear framework).

Figure 13 Different representations of the Keggin structure. left: polyhedral representation of the four $[W_3O_9]$ triads (yellow, red, blue, green) enclosing the tetrahedral template $[SiO_4]^{4-}$ (magenta). middle: View alongside one of the four Si-O bonds falling together with one of the eight C_3-axes. Top $[W_3O_9]$-triad (green) transparently colored showing the connectivity in ball and stick representation with the addenda atoms (blue), oxygen (red) and template (pink). right: Representation boiled down to the T_d-symmetry of the Keggin-cluster with the three atoms of each of the four triads interconnected and the templates Si-O bonds pointing towards the face of each triangle.

The monolacunary $[\alpha\text{-}SiW_{11}O_{39}]^{8-}$ is built exactly the same way only with one W-O unit missing, creating a vacancy of one polyhedron at its place (see Figure 14 below). It was chosen as the main precursor of this work mainly for the following three reasons: First of all is it a well investigated structure with established synthesis in literature ensuring reproducibility.[56, 59] Furthermore the heteropolyanions tungsten-oxygen framework is relatively flexible and can accommodate heteroelements of various size as template or as lacunary ion. Once formed, the substituted polyanions are remarkably resistant toward decomposition or rearrangement, if the right pH conditions are kept.[56] Beside that it is a relatively cheap starting material consisting of toxicological unproblematic elements which turns it into an ideal compound also with respect to potential applications.

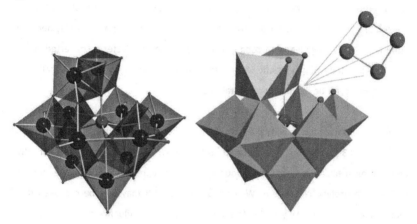

Figure 14 left: Ball and stick representation of the monolacunary Keggin cluster $[\alpha-SiW_{11}O_{39}]^{8-}$; color scheme: silicon (pink), tungsten (blue), oxygen (red). right: Polyhedral representation with $[WO_6]$-polyhedra (blue), heteroatom silicon (pink), oxygen (red) and highlighting of the square lacunary site with its four main oxo-ligands.

The precursor $K_8[\alpha-SiW_{11}O_{39}]\cdot 13H_2O$ is obtained from an aqueous solution of sodium tungstate (Na_2WO_4) that is slowly acidified with 4M HCl and combined with an aqueous solution of 1/11 equivalents of sodium metasilicate (Na_2SiO_3). The solutions pH is at this stage around 5 to 6 which initiates the aforementioned condensation mechanism and assembly into the cluster structure. Keeping the solution boiling for 1h leads to nearly quantitative conversion.

$$11 \, [WO_4]^{2-} + [SiO_3]^{2-} + 16 \, H^+ \xrightarrow[\text{boiling}]{1h} [\alpha\text{-}SiW_{11}O_{39}]^{8-} + 8 \, H_2O$$

Scheme 2 Reaction equation for the formation of the precursor $[\alpha\text{-}SiW_{11}O_{39}]^{8-}$ from an acidified solution containing silicate and tungstate in a stoichiometric molar ratio of 1:11.

Isolation of the cluster in form of its potassium salt is done by salting out the reaction mixture with KCl, prior to which the solution should be filtered or centrifuged once in order to remove solid traces of insoluble paratungstate. Washing the product with cool water removes remaining KCl. Nevertheless the product should be recrystallized once or twice from warm water to ensure that the precursor is as pure as possible, because impurities at this stage lead to wrong molar ratios when the cluster is converted into a ionic liquid in following synthetic steps. Work-up of the final POM-IL is practically not possible as it can´t be distilled or recrystallized.

An important factor for stability plays the pH that should be controlled with a precision of about 0.3 units on a calibrated pH meter. The potassium salt of the undecatungstosilicate is stable in aqueous solution between pH 4.5 to 7.0. When the pH increases, hydrolytic cleavage of W-O bonds occurs.[56] It may well be reasoned that mechanically first the $[SiW_{12}O_{40}]^{4-}$ is formed which upon the influence of the acidic pH is converted into its monolacunary polyanion $[SiW_{11}O_{39}]^{8-}$. Other well defined lacunary polyanions like the decatungstosilicate or nonatungstosilicate can be obtained in a similar fashion by choosing a pH in the right range.[56]

Interconversion of the dodecatungstosilicate and its various lacunary polyanions is possible and occurs readily. As these ions differ only slightly in their composition elemental analysis is not a reliable criterion for purity and it is more convenient to characterize the product by its physiochemical properties like its vibrational spectrum (see chapter 3.1.3.1).

Incorporation of a d-element into the void of the flexible framework of the $\{W_{11}\}$ cluster is relatively straight forward. The lacuna may accommodate cations of widely varying size ranging from 53 pm (Al^{3+}) to at least 88 pm (Tl^{3+}) ionic radius [59] which is an important prerequisite for the aim of this work to vary the lacunary heteroatom. The lacuna behaves like a pentadentate ligand with the four outer oxo ligands binding to the incorporated d-metal together with one elongated bond that is formed to the nearest oxo ligand of the $[MO_4]$ template.

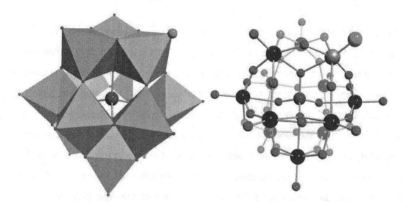

Figure 15 left: Polyhedral representation of the d-element substituted Keggin cluster $K_x[\alpha\text{-}SiW_{11}O_{39}M(H_2O)]$; templating silicon (magenta), addenda atoms tungsten (blue), oxygen (red), transition metal M (grey) and a sixth ligand (e.g. representing H_2O) coordinated to M (green). right: The same cluster in ball and stick representation.

As a general synthetic procedure the following approach was applied for any kind of d-metal (M) applied in this study:

In a first synthetic step $K_8[\alpha\text{-}SiW_{11}O_{39}]\cdot13H_2O$ is suspended in water, heated to 50-60 °C and an equimolar solution of the d-element in form of its halide or nitrate salt added drop wise. The solution is stirred for one hour and the pH acidified to the right degree (for details see the experimental part). In the following example the reaction is exemplified for the corresponding copper compound:

$$K_8[\alpha\text{-}SiW_{11}O_{39}]\cdot13H_2O + CuCl_2 \xrightarrow[\text{HCl}]{\text{1h}} K_6[\alpha\text{-}SiW_{11}O_{39}Cu(H_2O)]\cdot nH_2O + 2\ KCl$$

Scheme 3 Reaction (I): incorporation of a d-element (exemplified for Cu^{II}) into the monovacant Keggin-cluster $[SiW_{11}O_{39}]^{8-}$.

Altogether seven different d-element ions (M) were build into the Keggin cluster with $M = Cu^{II}, Co^{II}, Fe^{III}, Ni^{II}, Fe^{II}, Cr^{III}, Mn^{II}$.

3.1.2 Phase transfer of the Si-templated Keggin clusters

Phase transfer follows subsequently after cooling the reaction mixture to room temperature. The transfer of clusters into an organic aprotic solvent by utilizing their respective tetrabutylammonium salts is actually known for some time. Their synthesis has been pioneered by *Jahr, Fuchs* and *Oberhauser* who first prepared salts of $[Mo_6O_{19}]^{2-}$, $[W_6O_{19}]^{2-}$ and $[W_{10}O_{32}]^{4-}$ by controlled hydrolysis of molybdic and tungstic acid esters.[60, 61] In the work at hand the phase transfer was achieved in a more convenient way by mixing of an aqueous solution of the polyoxometalate with the tetraalkylammonium-bromide dissolved in toluene. Extraction could be realized directly out of the reaction mixture for Si^{4+}-templated clusters of the type {$SiW_{11}M$} whose preparation according to Scheme 3 is practically quantitative.

The tetraalkylammoniumbromide (Q-Br) phase transfer reagent is dissolved in toluene and added with a molar ratio Q-Br : $K_x[SiW_{11}O_{39}M(H_2O)]$ of almost x : 1. A very slight deficit of around 0.1 equivalents delivered cleaner reaction products according to EA, which may reasonably be explained by slight amounts of insoluble paratungstate formed from the precursor which is not available for phase transfer with the Q^+. So in case of the copper compound 5.9 molar equivalents of Q-Br were added for complete substitution of all K^+ counterions.

$$K_6[\alpha\text{-}SiW_{11}O_{39}Cu(H_2O)]_{(aq)} + 6\ Q\text{-}Br_{(tol)} \longrightarrow (Q)_6[\alpha\text{-}SiW_{11}O_{39}Cu(H_2O)]_{(tol)} + 6\ KBr_{(aq)}$$

Scheme 4 Chemical reaction equation for the phase transfer of the POM obtained from reaction (I) into toluene using Q-Br exemplified for $K_6[SiW_{11}O_{39}Cu(H_2O)]\cdot H_2O$. (aq)-Indices represent species dissolved in the aqueous phase, (tol)-indices represent species in the organic toluene phase of the biphasic reaction mixture.

The product is isolated afterwards in the organic layer with a separation funnel. For removal of residual water the crude product is solvent stripped with toluene followed by chloroform. Finally it is dried for at least two days under vacuum with several times lyophilizing and successive heating to 60°C.

The following compounds were obtained following the same reaction procedure: $(Q^5)\{SiW_{11}Cu\}$ (1), $(Q^5)\{SiW_{11}Fe\}$ (2), $(Q^6)\{SiW_{11}Cu\}$ (3), $(Q^6)\{SiW_{11}Fe\}$ (4), $(Q^7)\{SiW_{11}\}$ (5), $(Q^7)\{SiW_{11}Cu\}$ (6), $(Q^7)\{SiW_{11}Co\}$ (7), $(Q^7)\{SiW_{11}Fe^{III}\}$ (8), $(Q^7)\{SiW_{11}Ni\}$ (9), $(Q^7)\{SiW_{11}Cr\}$ (10), $(Q^7)\{SiW_{11}Fe^{II}\}$ (11), $(Q^7)\{SiW_{11}Mn\}$ (12), $(Q^8)\{SiW_{11}Cu\}$ (13), $(Q^8)\{SiW_{11}Fe\}$ (14)

For the Q^7-series iron was applied in its oxidation state of +II (11) and +III (8). Over a period of several days the yellow colored compound (11) turned the same color as its brown colored Fe^{III} analogue (8), which indicates an oxidation of the iron (II) to iron (III) (by oxygen from air or the cluster itself). The compounds IR and UV-Vis spectra were identical at this point and both compounds showed the same EPR-signal. As air-stable ionic liquids were the main goal of this work, no further investigation of compound (11) were undertaken which would have required glove box atmosphere or stabilizing agents for the Fe(II) oxidation state.

3.1.3 Analysis of the synthesized Keggin clusters
3.1.3.1 Vibrational spectroscopy

The Keggin structure of the anions was evidenced by IR spectroscopy on a Shimadzu FT-IR-8400S spectrometer by assignment of characteristic vibrations as they are listed in literature.[62]

assignment	TPA		THexA		THA		TOA	
	SiW$_{11}$Cu	SiW$_{11}$Fe	SiW$_{11}$Cu	SiW$_{11}$Fe	SiW$_{11}$Cu	SiW$_{11}$Fe	SiW$_{11}$Cu	SiW$_{11}$Fe
$\upsilon_{as}(X-O_c)+\upsilon_{as}(W-O_t)$	1001 w	1001 w	1001 w	1002 w	1009 w	1001 w	1006 w	1003 w
$\upsilon_{as}(W-O_t)$	951 m	955 m	951 m	958 m	962 m	957 m	953 m	959 m
$\upsilon_{as}(W-O_c) + \upsilon_{as}(W-O_t)$sym	907 vs	910 vs	908 vs	912 vs	908 s	914 s	907 s	914 vs
$\upsilon_{as}(W-O_v-W)$	-	876 sh	-	878 m	-	879 m	-	878 m
$\upsilon_{as}(W-O_a-W) + \upsilon_{as}(Cu-O)^a$	802 vs	798 vs	804 vs	799 s	802	799 s	800 s	799 vs
$\upsilon_{as}(Cu-O)$	689 m	-	692 m	-	689 m	-	689 w	-
$\delta(W-O_a-W) + \delta(M-O_c)$	540 w	527w	540 w	525 w	540 w	527 w	540 w	526 w
$\delta(W-O_b-W) + \delta(M-O_c)$	525 w	505 w	529 w	500 w	524 w	506	529 w	505
$\delta(W-O_b-W)$	414 w	418 w	419 w	419 w	417 w	-	417 w	419 w

Table 2 Assignment of characteristic IR vibrations for the compounds $(Q^V)_x[SiW_{11}O_{39}M(H_2O)]$ with y = 5-8 and M = Cu^{II} and Fe^{III}.

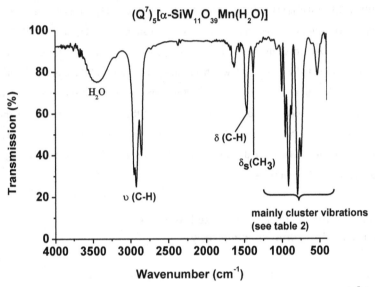

Figure 16 Exemplary IR spectrum for the compounds of tetraalkylammonium Keggin samples: $(Q^7)_6[\alpha$-$SiW_{11}O_{39}Mn(H_2O)]$ obtained in KBr-pellet.

Around 3500 cm^{-1} a broad peak is observed for all studied samples arising due to water bound as ligand at the sixth coordination position of the d-element heteroatom. Its broadness and position at a relatively high wave number are indicators for strong hydrogen bonds the water ligand can form to adjacent cluster oxo-atoms. The quaternary ammonium cations are readily evidenced by their characteristic C-H-valence vibrations (around 2960-2850 cm^{-1}), deformation vibrations (around 1460-1480 cm^{-1}) and at around 1380 cm^{-1} for the symmetric deformation of the CH$_3$ group.[63] The CH$_2$ rocking vibration at around 720 cm^{-1} is relatively weak and may in some cases be overlaid by the combined υas(W-O$_a$-W) + υas(C$_u$-O)$_b$ valence vibration of the cluster with its centre at around 750 cm^{-1}.

3.1.3.2 UV-Vis spectroscopy

UV–vis spectra were recorded on a Shimadzu UV-2401PC spectrophotometer in transmission mode with quartz cuvettes (1.0 cm optical path length). Samples **(1)-(4)** and **(11)+(12)** were recorded in CH_2Cl_2, samples **(5)-(10)** were recorded in toluene as solvent. The Keggin cluster dominates the absorption features and shows characteristic absorption peaks at around 228 and 258 nm which are assigned to the intramolecular ($O_d \rightarrow$ W and $O_{b,c} \rightarrow$ W) ligand to metal charge transfer (LMCT) transitions.[64, 65] In addition absorption peaks in the visible spectrum occur for all clusters except for iron due to the incorporated hetero-element which is sixfold coordinated to oxygen (five oxo ligands by the Keggin cluster and one oxygen bound water ligand) which suggests a comparison with the well known hexaqua complexes of the respective heteroelement. In case of the iron(III) d^5-compounds, where d $\rightarrow$ d excitations are spin forbidden, the clusters yellow color is caused by a batochromic shift of the LMCT absorption due to the irons charge of +3 (compared e.g. to the isoelectronic Mn(II) analogues).[66]

The assignment of all peaks was done with the help of similar data from literature (see footnote of Table 3 or with the corresponding Tanabe-Sugano diagrams, which were particularly useful for additionally assigning spin-forbidden intercombination transitions where they occurred. A detailed overview of the UV-Vis data is given below in Table 3:

No.	short-formula	concentration	absorption wave-length (nm)	absorption wave-number (cm⁻¹)	extinction coefficient ($l \cdot mol^{-1} \cdot cm^{-1}$)	assignment	absorption wave-number of corresponding hexaquaion (cm⁻¹)
1	$(Q^5)_6\{SiW_{11}Cu\}$	18.0 μM	228	43860	41446.3	LMCT	-
			258	38760	41581.4	LMCT	-
		18.0 mM	715	13986	23.1	$^2E_g \rightarrow {}^2T_{2g}$	12500[a]
			880	11364	9.3	$^2E_g \rightarrow {}^2T_{2g}$	-
2	$(Q^5)_5\{SiW_{11}Fe\}$	28.5 μM	229	43668	26808.3	LMCT	-
			263	38023	26100.8	LMCT	-
3	$(Q^6)_6\{SiW_{11}Cu\}$	35.3 μM	229	43668	27748.6	LMCT	-
			257	38911	26080.3	LMCT	-
		17.6 mM	715	13986	22.4	$^2E_g \rightarrow {}^2T_{2g}$	12500[a]
			870	11494	11.3	$^2E_g \rightarrow {}^2T_{2g}$	-
4	$(Q^6)_5\{SiW_{11}Fe\}$	35.3 μM	229	43668	27748.6	LMCT	-
			257	38911	26080.3	LMCT	-
6	$(Q^7)_6\{SiW_{11}Cu\}$	45.0 μM	283	35336	16165.2	LMCT	-
		17.9 mM	710	14085	18.6	$^2E_g \rightarrow {}^2T_{2g}$	12500[a]
						$^2E_g \rightarrow {}^2T_{2g}$	-
7	$(Q^7)_6\{SiW_{11}Co\}$	23.0 μM	284	35211	36731.0	LMCT	8700[a]
		3.9 mM	563	17762	67.8	$^4T_{1g} \rightarrow {}^4T_{2g}$	16000[a]
			520	19231	56.9	$^4T_{1g} \rightarrow {}^4T_{2g}$	19400[a]
8	$(Q^7)_5\{SiW_{11}Fe\}$	65.0 μM	283	35336	21636	LMCT	-
9	$(Q^7)_6\{SiW_{11}Ni\}$	64.0 μM	285	35088	15704	LMCT	-
		10.4 mM	432	23148	26.1	$^3A_{2g} \rightarrow {}^3T_{1g}$	25300[a]
			740	13516	5.8	$^3A_{2g} \rightarrow {}^1E_g$	-
			806	12407	7.5	$^3A_{2g} \rightarrow {}^3T_{2g}$	13800[a]
						$^3A_{2g} \rightarrow {}^3T_{2g}$	8500[a]
10	$(Q^7)_5\{SiW_{11}Cr\}$	37.0 μM	284	35211	26012	LMCT	-
		20.4 mM	646	15480	31.3	$^4A_{2g} \rightarrow {}^4T_{2g}$	17400[a]
			717	13947		$^4A_{2g} \rightarrow {}^2E_{2g}$	24500[a]
						$^4A_{2g} \rightarrow {}^4T_{1g}$	38600[a]
12	$(Q^7)_6\{SiW_{11}Mn\}$	27.0 μM	284	35211	19819	LMCT	18600[b]
		1.62 mM	507	19724	263.7	$^3T_1 \rightarrow {}^3A_2$	22900[b]
			522	19157	278.4	$^3T_1 \rightarrow {}^3A_1$	24900[b]
			538	18587	289.4	$^3T_1 \rightarrow {}^3T_2$	25150[b]
			575	17391	180.3	$^3T_1 \rightarrow {}^3E$	
13	$(Q^8)_6\{SiW_{11}Cu\}$	18.2 μM	259	38610	30145.2	LMCT	-
		18.2 mM	720	13889	28.8	$^2E_g \rightarrow {}^2T_{2g}$	12500[a]
			910	10989	7.8	$^2E_g \rightarrow {}^2T_{2g}$	-
14	$(Q^8)_5\{SiW_{11}Fe\}$	9.3 μM	259	38610	58976.3	LMCT	-

a) Hollemann Wiberg Lehrbuch der Anorganichen Chemie,102.Auflage, 2007, de Gruyter, p.1371
b) Hollemann Wiberg Lehrbuch der Anorganichen Chemie,102.Auflage, 2007, de Gruyter, p.1613

Table 3 UV-Vis spectroscopic analysis of compounds 1-12, giving absorption wavelengths and frequencies, extinction coefficients and transition assignments. Absorption wavenumbers for the corresponding hexaqua ions are given for comparison

An evaluation of the obtained absorption wave numbers shows that the coordination of d-elements M in the lacuna of the monolacunary Keggin cluster induces a hypsochromic shift. The cluster acts as a pentadentate chelating ligand and stabilizes the complexes non-bonding n-orbitals as well as the antibonding π^*-orbitalts which are responsible for the absorption characteristics. As the n-orbitals are stronger lowered in energy than the π^*-orbitalts an increased excitation energy ΔE is the consequence leading to absorptions of lower wavelengths (or higher wavenumbers) according to $\lambda = \frac{h \cdot c}{\Delta E}$.

Overall a striking similarity between the absorption spectra of the POM-ILs investigated and the respective hexaquaions in aqueous solution exists, which is a good indicator that the octahedral coordination environment of the d-metal heteroelement M is retained in the organic phase with an additional water ligand bound in the sixth coordination position. The applicability of the Tanabe-Sugano diagrams is a further proof for this.

A detailed analyses of the UV-Vis spectra may also be used to gain clarity of the heterometals oxidation state after its incorporation into the cluster. Particularly the strongly colored $(Q^7)_6[\alpha\text{-SiW}_{11}O_{39}Mn(H_2O)]$ that was synthesized with $MnCl_2 \cdot 4H_2O$ is an interesting example as it seems to undergo a change of its oxidation state from Mn(II) d^5 to Mn(III) d^4, which is reflected in the appearance of four peaks in its UV-Vis spectrum that can only be interpreted at hand of the Tanabe Sugano diagram for a d^4 low-spin configuration (see Figure 17).

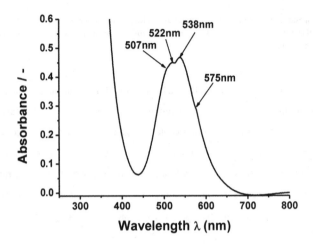

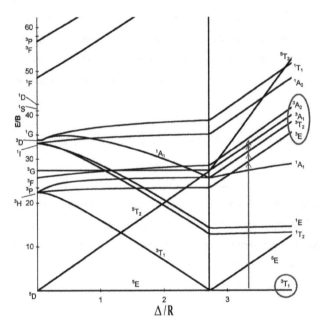

Figure 17 *top*: UV-Vis spectrum of $(Q^7)_6[\alpha\text{-SiW}_{11}O_{39}Mn(H_2O)]$ in toluene (c = 1.62 mM) with peak positions at wavelengths of λ = 507, 522, 538 and 575 nm. *bottom*: Tanabe-Sugano diagram for a d^4 electron configuration. In the low-spin state four excitations from the triplet ground state $^3T_1$ are possible, namely into ^3E, $^3T_2$, $^3A_1$ and $^3A_2$.

3.1.3.3 Thermogravimetric analysis

The thermal stability of **1 – 12** was examined by thermogravimetric analysis under helium atmosphere in a temperature range from room temperature to 800 °C with a heating rate of 5 K/min. As all compounds showed similar thermogravimetric behavior, the analysis is exemplified using compound **5**, $(Q^7)_6[\alpha\text{-SiW}_{11}O_{39}Cu(H_2O)]$ (see Figure 18). In a first step (T < 200-210 °C), residual solvent and cluster-based water-ligands are lost (weight loss between *ca.* 1-11%). At higher temperatures, decomposition of the organic cation is observed until a temperature of *ca.* 400-450 °C is reached, where all organic materials have been fully decomposed. For detailed analyses, see Table 4 (page 43).

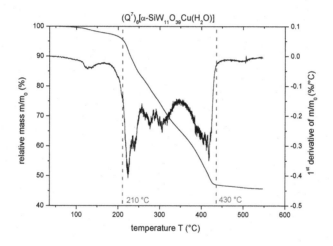

Figure 18: TGA of $(Q^7)_6[\alpha\text{-SiW}_{11}O_{39}Cu(H_2O)]$ obtained under He atmosphere with a heating rate of 5 °C/min. Relative mass m/m_0 as function of temperature (blue line) and its first derivative (black line).

When analyzing the thermal stability, the following trend was observed: Increase of the alkyl chain length of the cations results in a slight decrease of thermal stability. This is exemplified when comparing the Q^5-based compounds (decomposition onset: *ca.* 200/210 °C) with the Q^8-based compounds (decomposition onset: *ca.* 190/195 °C). When comparing the influence of different metal substitution patterns (e.g. Cu^{II} vs Fe^{III}-

substitution) for identical cations, only marginal changes in thermal stability are observed.

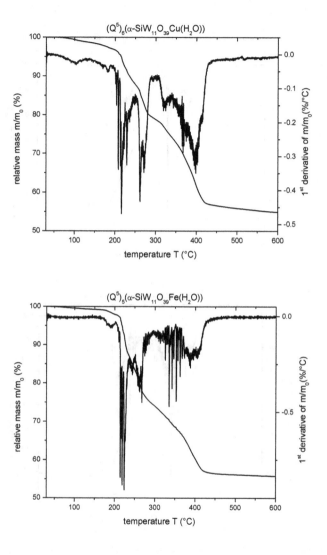

Figure 19 TGA of the Keggin clusters charge balanced with Q^5. TGA obtained under He atmosphere with a heating rate of 5 °C/min.

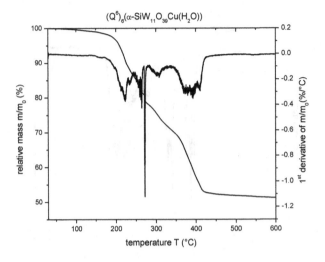

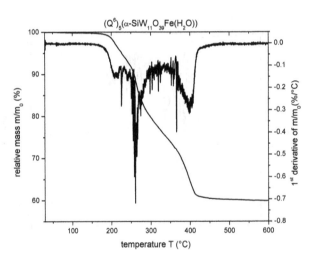

Figure 20 TGA of the Keggin clusters charge balanced with Q^6. TGA obtained under He atmosphere with a heating rate of 5 °C/min.

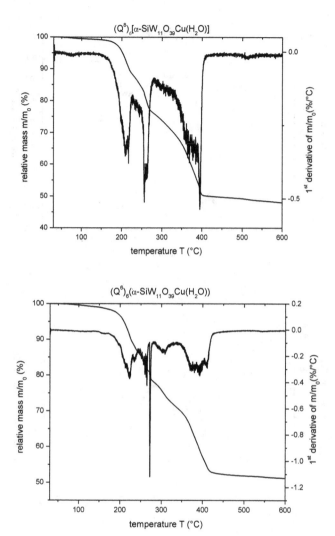

Figure 21 TGA of the Q⁸ based ILs obtained under He atmosphere with a heating rate of 5 °C/min

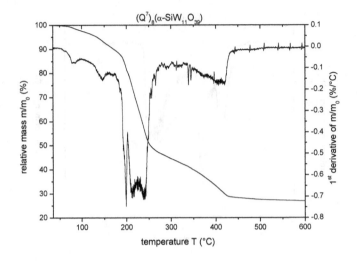

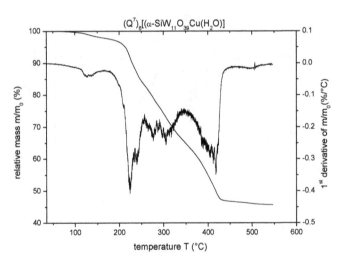

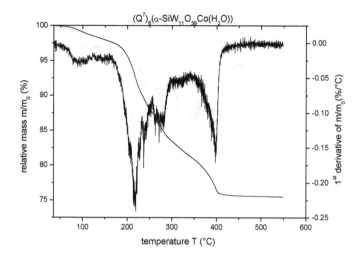

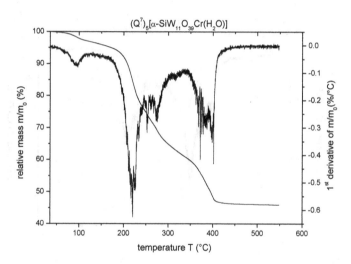

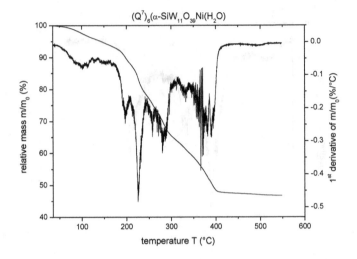

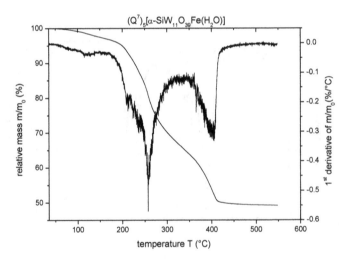

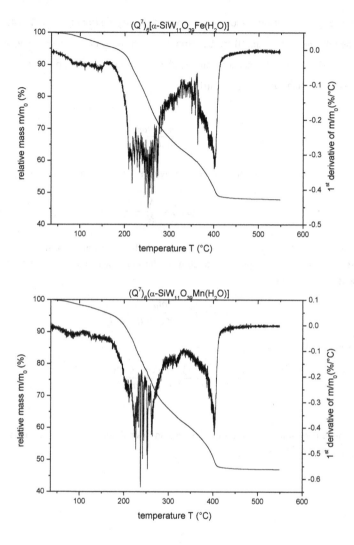

Figure 22 TGA of the Q^7 based ILs obtained under He atmosphere with a heating rate of 5 °C/min.

As the second step of weight loss in the temperature regime from ca. 190 °C to 450 °C (in the following called $\Delta m(II)$) can be attributed to the decomposition of the organic cation can it be utilized to calculate the stoichiometric number of cations N in the molecular formula $(Q^Y)_N[SiW_{11}O_{39}M(H_2O)]$.

$$\text{It is: } N = \frac{n(Q^y\ lost)}{n(cluster)}$$

$n(Q^Y$ lost) can be directly calculated from the weight loss $\Delta m(II)$ according to

$$n(Q^y\ lost) = \frac{\Delta m(II)}{M(Q^y)}$$

For calculation of n(cluster) its correct mass has to be determined given as the samples mass m_0 which was weight in subtracted by the mass of residual solvent of the first weight loss step $\Delta m(I)$.

$$n(cluster) = \frac{m_0 - \Delta m(I)}{M(compound)}$$

So N was calculated as:

$$N = \frac{\Delta m(II) \cdot M(compound)}{M(Q^y)(m_0 - \Delta m(I))}$$

A detailed overview of the measured values and calculated results is given on the next page in Table 4.

43

entry No.	formula	T range of Δm (II) (°C)	m_0 of sample (mg)	Δm (I) (%)	Δm (II) (%)	N (cations) calculated
1	$(Q^5)_6[\alpha\text{-SiW}_{11}O_{39}Cu(H_2O)]$	200-420	53.539	3.7	39.1	6.16
2	$(Q^5)_5[\alpha\text{-SiW}_{11}O_{39}Fe(H_2O)]$	210-420	25.542	3.9	38.1	5.61
3	$(Q^6)_6[\alpha\text{-SiW}_{11}O_{39}Cu(H_2O)]$	200-415	61.617	2.9	44.1	6.23
4	$(Q^6)_5[\alpha\text{-SiW}_{11}O_{39}Fe(H_2O)]$	205-415	33.474	1.3	37.3	5.08
5	$(Q^7)_6[\alpha\text{-SiW}_{11}O_{39}Cu(H_2O)]$	210-430	74.235	4.5	46.5	6.19
6	$(Q^7)_6[\alpha\text{-SiW}_{11}O_{39}Co(H_2O)]$	195-405	40.279	2.6	43.5	6.18
7	$(Q^7)_6[\alpha\text{-SiW}_{11}O_{39}Ni(H_2O)]$	200-405	72.857	11.0	41.4	5.91
8	$(Q^7)_6[\alpha\text{-SiW}_{11}O_{39}Mn(H_2O)]$	200-410	59.270	6.8	45.0	6.13
9	$(Q^7)_5[\alpha\text{-SiW}_{11}O_{39}Fe(H_2O)]$	205-410	75.805	6.4	43.1	5.38
10	$(Q^7)_5[\alpha\text{-SiW}_{11}O_{39}Cr(H_2O)]$	200-405	19.517	8.6	43.5	5.56
11	$(Q^8)_6[\alpha\text{-SiW}_{11}O_{39}Cu(H_2O)]$	190-400	29.210	2.4	46.4	6.14
12	$(Q^8)_5[\alpha\text{-SiW}_{11}O_{39}Fe(H_2O)]$	195-400	31.631	3.7	44.3	5.01

Table 4 Thermogravimetric analysis for compounds 1-12, showing the mass loss steps and number of cations observed.

From the calculated amount of cations their stoichiometric number and therefore the successful cluster synthesis and its practically quantitative phase transfer is confirmed. All calculated values N show slightly higher values than their respective stoichiometric number because small amounts of residual solvent may still contribute to the weight loss at higher temperatures (e.g. cluster bound water; see also the discussion of IR-data in chapter 3.1.3.1). Additionally, small amounts of the thermally stable cluster can be carried away under the influence of heat and the gas stream, also adding weight loss to Δm(II) additionally to the loss of Q^y, why slightly higher values are not only reasonable but to be expected for the calculated number of counterions N.

3.1.4 Physiochemical properties of the Si-templated Keggin clusters

3.1.4.1 Miscibility properties

All the investigated ionic liquids are miscible with typical organic solvents having low, medium and high dielectric constants (ε) like toluene (ε = 2.38), chloroform (ε = 4.81), acetone (ε = 20.70), ethanol (ε = 24.55), methanol (ε = 32.70), acetonitrile (ε = 37.50), DMSO (ε = 46.68) and so forth. Their solubility in those media increases strongly with rising temperature. Immiscibility is only observed with water, also at higher temperatures. Most of the known ionic liquids are miscible with organic solvents and with water. The present compounds might therefore be interesting candidates for new biphasic synthesis, catalysis and separation.

3.1.4.2 Conductivity

The electrical conductivity σ as a function of temperature was exemplary investigated using the model system $(Q^7)_6[\alpha\text{-SiW}_{11}O_{39}Cu(H_2O)]$. As expected, the electric conductivity increases exponentially with temperature, reaching values of σ *ca.* 5 $\mu S \cdot cm^{-1}$ at T = 85 °C, see Figure 23. The relatively low conductivity is most likely associated with the high viscosity of the POM-ILs which decreases with temperature. For classic ionic liquids typical conductivities of σ < 10 – 15 $mS \cdot cm^{-1}$ are reported.[9]

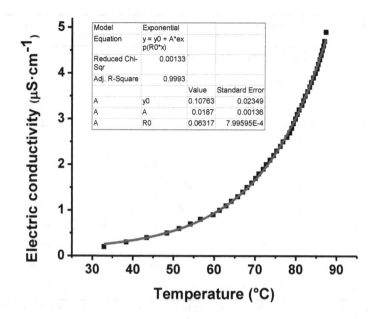

Figure 23. Electric conductivity as a function of temperature (T = 35 °C – 88 °C) for the ionic liquid $(Q^7)_6[\alpha\text{-SiW}_{11}O_{39}Cu(H_2O)]$ (black squares) and exponential fit (red line).

3.1.4.3 Melting ranges

All of the Keggin clusters synthesized yielded an ionic liquid with Q^7 or Q^8 as the counterion and with d-elements having a two (e.g. Cu^{II}, Mn^{II}) or three times (e.g. Cr^{III}, Fe^{III}) positive charge. With alkyl chains on the tetraalkylammonium ions shorter than seven carbon atoms (Q^n n < 7) the product was obtained as a solid at room temperature. Table 5 gives a general overview of the charges and the products melting ranges.

Compound no	compound formula	metal salt used	charge of cluster anion	melting range (°C)
1	$(Q^5)_6[\alpha\text{-}SiW_{11}O_{39}Cu(H_2O)]$	$Cu(NO_3)_2\cdot 3H_2O$	6-	200-210
2	$(Q^5)_5[\alpha\text{-}SiW_{11}O_{39}Fe(H_2O)]$	$FeCl_3\cdot 6H_2O$	5-	220-228
3	$(Q^6)_6[\alpha\text{-}SiW_{11}O_{39}Cu(H_2O)]$	$Cu(NO_3)_2\cdot 3H_2O$	6-	120-130
4	$(Q^6)_5[\alpha\text{-}SiW_{11}O_{39}Fe(H_2O)]$	$FeCl_3\cdot 6H_2O$	5-	78-90
6	$(Q^7)_6[\alpha\text{-}SiW_{11}O_{39}Cu(H_2O)]$	$Cu(NO_3)_2\cdot 3H_2O$	6-	RT-IL
7	$(Q^7)_6[\alpha\text{-}SiW_{11}O_{39}Co(H_2O)]$	$Co(NO_3)_2\cdot 6H_2O$	6-	RT-IL
8	$(Q^7)_5[\alpha\text{-}SiW_{11}O_{39}Fe(H_2O)]$	$FeCl_3\cdot 6H_2O$	5-	RT-IL
9	$(Q^7)_6[\alpha\text{-}SiW_{11}O_{39}Ni(H_2O)]$	$NiCl_2\cdot 6H_2O$	6-	RT-IL
10	$(Q^7)_5[\alpha\text{-}SiW_{11}O_{39}Cr(H_2O)]$	$Cr(NO_3)_3\cdot 9H_2O$	5-	RT-IL
12	$(Q^7)_6[\alpha\text{-}SiW_{11}O_{39}Mn(H_2O)]$	$MnCl_2\cdot 4H_2O$	6-	RT-IL
13	$(Q^8)_6[\alpha\text{-}SiW_{11}O_{39}Cu(H_2O)]$	$CuCl_2\cdot 2H_2O$	6-	RT-IL
14	$(Q^8)_5[\alpha\text{-}SiW_{11}O_{39}Fe(H_2O)]$	$FeCl_3\cdot 6H_2O$	5-	RT-IL

Table 5 Overview of all compounds of the type $(Q^x)_n[\alpha\text{-}XW_{11}O_{39}TM(H_2O)]$ showing the transitionmetals used for derivatizing the lacuna, the cluster charges and melting ranges.

A comparison of the melting ranges shows that longer alkyl chains result in lower melting points which is nicely seen at hand of the $[\alpha\text{-}SiW_{11}O_{39}Cu(H_2O)]^{6-}$-based compounds **1**, **3**, **6**, **13** where the melting range decreases as the cation alkyl chain length increases from Q^5 (m.p. 200-210 °C) to Q^8 (RTIL). The behavior can be explained by the increasing steric demand of the longer chains. This results in lower electrostatic interactions between cation and anion, due to increased cation-anion distances. As a result, lower melting points are observed for longer alkyl chains. When studying the influence of the cluster charge and therefore, the number of counter cations, an interesting trend is observed: a higher number of counter ions results in a decreased melting point. The melting point of compound **1** (cluster charge: 6-) is approximately 20 °C lower than the melting point of **2** (cluster charge: 5-). The effect is increased with increasing cation size: while the difference in melting point for the Q^5-based compounds **1** and **2** is *ca.* 20 °C, the related, Q^6-based compounds **3** and **4** feature a melting point difference of *ca.* 40 °C.

In order to determine the melting points and gain insight into the melting behavior of the ionic liquid products differential scanning calorimetry (DSC) was used. The investigation of the Q^7-based POM-ILs revealed notable differences between the samples. It was observed that the RTILs **6, 8** and **9** feature a well defined melting peak with an onset temperature of around 24 °C in combination with a minor eutectic melt located at around -16 °C. In contrast, **7, 10** and **12** show a broad asymmetric melting range between ca. -70 °C and 0 °C with a minimum located at ca. -50 °C to -60 °C. It is proposed that the difference in melting behavior is due to the transition metal cations incorporated in the cluster shells: compounds **6, 8** and **9** feature metal cations which are redox-inert under the given reaction conditions (Cu^{II}, Ni^{II}, Fe^{III}, respectively). In contrast, compounds **7, 10** and **12** contain redox-active metals (Mn^{II}, Co^{II}, Cr^{III}) and it is proposed that under the given conditions, partial oxidation results in the in-situ generation of "intrinsic" impurities which contribute to the unexpected melting behavior.

3.1.4.4   Rheological analysis

Rheology is an excellent tool to describe how matter is flowing, primarily in the liquid state, but also as soft matter or solids under conditions in which they respond with plastic flow to an applied external force. It generally accounts for the behavior of non-Newtonian fluids that show a strain rate dependent viscosity. The relative movement of different layers in a given material can cause a change of its viscosity, which may be reduced upon strain (so called shear thinning materials) or which may even rise with relative deformation (so called shear thickening or dilatant materials). The experimental characterization of a material's rheological behavior (so called rheometry) is carried out by dynamic mechanical analysis with an oscillatory force (stress) being applied to the material and the resulting displacement (strain) being measured. In purely elastic materials the stress and strain occur in phase, so that the response of one occurs simultaneously with the other. By contrast in purely viscous materials, there is a phase difference between stress and strain, where strain lags by a 90° phase lag. Most materials exhibit a behavior somewhere in between with a phase lag in strain between 0° and 90° - so called viscoelastic materials.

For the representation of stress and strain in a viscoelastic material the following sinusoidal expressions can be used:

strain: $\qquad \varepsilon = \varepsilon_0 \sin(\omega t)$ (I)

stress: $\qquad \sigma = \sigma_0 \sin(\omega t + \delta)$ (II)

where $\omega = 2\pi f$ with f being the frequency of strain oscillation, t is the time and δ is the phase lag between stress and strain.

At hand of the so called modulus G* the stiffness of a material can be described. It can be divided into two parts, the storage modulus G' which describes the elastic response (stored energy), and G", the loss modulus which describes the viscous properties (energy dissipated as heat). The shear storage and loss moduli are defined as follows:

storage modulus $\qquad G' = \dfrac{\sigma_0}{\varepsilon_0} \cos \delta$ (III)

loss modulus $\qquad G'' = \dfrac{\sigma_0}{\varepsilon_0} \sin \delta$ (IV)

The sinusoidal stress σ is dissipated in the sum of both moduli:

$$\sigma = \varepsilon_0 (G'_{(\omega)} \sin \omega t + G''_{(\omega)} \cos \omega t)$$ (V)

The moduli can be measured as shown in Figure 24 (see next page). A viscoelastic material is poured into a cup and a concentrical cylinder (called "bob") is lowered into the material. The cup is oscillating in a sinus like movement and the force which the material is transferring is measured with the concentric cylinder. The force will also be sinus like with the same frequency as the cup, but displaced in comparison to the movement. The displacement is measured as a phase angel δ. A totally elastic material, such as steel, has no displacement at all, and the phase angel $\delta = 0°$, while a ideal liquid has maximal displacement and $\delta = 90°$. The onset deformation, the measured force and displacement are used to describe the storage modulus and the loss modulus. Viscoelastic measurements at different frequencies can be used to describe the character for a material in the same way as an optical spectrum is used for spectrophotometric characterization of compounds.

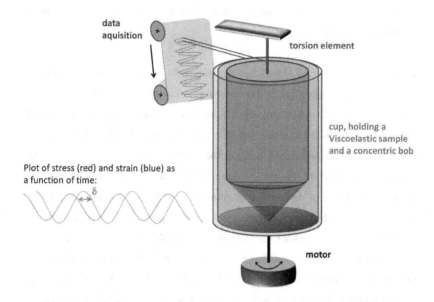

Figure 24 Sketch of a rheometer for viscoelastic measurements. The viscoelastic sample is placed in a cup and exposed to an external sinusoidal stress. This is plotted together with the strain response as a function of time in order to determine the phase lag δ (see green double arrow).

The rheological properties of ionic liquids in general are influenced by the degree of electrostatic interaction between the ILs cation and anion.[67, 68] POM-ILs in particular show a great change in their viscosity at room temperature because this temperature region is close to their melting point. Furthermore the molecular volumes of ILs are correlated with their viscosity and their electrical conductivity.[69] The viscosity of inorganic ionic liquids like POM-ILs mainly depends on weak electrostatic interaction forces. *Dai* and coworkers described a class of ionic liquids formed by titanium tamplated polyoxometalates of the Keggin-type with sodium counterions.[69] These ionic liquids show low viscosities ranging from 42 to 82 mPa·s. Their findings need however to be treated critically as the investigated cluster as they describe them should not be able to bind Lanthanides. Further, their IL requires 10 percent or more water present, otherwise it is converted into a solid, which means that the liquid state at room temperature is not an inherent feature of the compound but stemming from colligative effects with the water molecules. One further (and more reliable) reference in current literature is that of *Rickert et al.* who investigated Keggin polyoxometalates charge balanced with tetraalkylphosphonium cations.[67] These show a much higher viscosity at room temperature of η = 75000 mPa s.

Within the measured range of the amplitude sweeps the compounds No. **6-11** are linear viscoelastic liquids suggesting that the internal structure of the ionic liquid is stable.[70] Compounds **10** and **12** show almost identical loss modulus values; for compounds **6, 7, 8** and **9**, the loss modulus is approximately one order of magnitude larger, demonstrating that cation variation can be used to modify the POM-IL rheology. In contrast, the linear viscoelastic regime of the Q^8-based compounds **13** and **14** is restricted with respect to the accessible range of deformation (deformation < 1 %). In particular, the storage modulus of **13** and **14** decreases considerably, indicating significant structural changes within the IL. As this behavior was only observed for Q^8-based materials, a possible explanation based on the size-difference between Q^7 and Q^8 cations is proposed: the average cation-anion distance in the Q^8-based POM-ILs **13** and **14** is expected to be larger compared with the Q^7-based POM-ILs **6-9+12**. As a result, the attractive electrostatic interactions in **13** and **14** will be smaller and large deformations cannot be compensated, resulting in the observed decrease of the storage modulus and the related structural changes.

The frequency sweep tests for compounds No. **6-12** in the linear viscoelastic range showed very good reproducibility, indicating high sample stability. Storage and loss moduli increase continuously with frequency close to a typical Maxwell fluid in the low frequency limit where the storage modulus is proportional to the square of the frequency, the loss modulus increases linearly with the frequency and the complex viscosity is constant. Frequency sweeps for compounds **13** and **14** were carried out in their linear viscoelastic regimes at deformation of 0.05% and 0.5%, respectively. Here, loss and in particular storage moduli increase with exponents lower than one suggesting non-Maxwell fluid behavior. For **14**, both moduli show almost identically frequency dependency.

A)

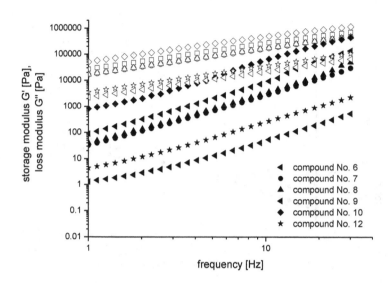

B)

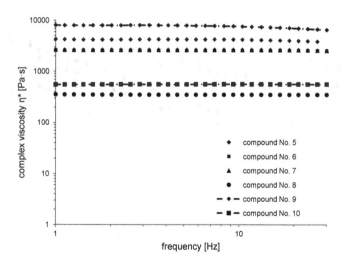

C)

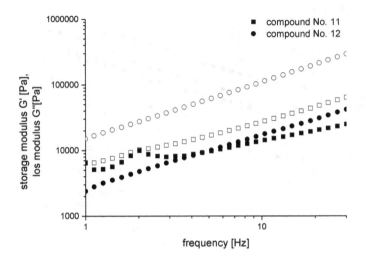

D)

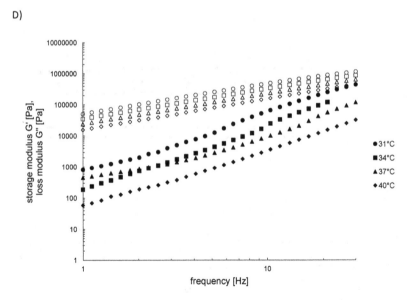

Figure 25 (A) Frequency sweep and complex viscosity with loss modulus (hollow symbols) and storage modulus (open symbols). (B) Frequency sweep for compounds 6-12 at a strain amplitude of 5%. Sample $(Q^7)_5[\alpha\text{-SiW}_{11}O_{39}Fe(H_2O)]$ (**8**) was measured at T = 31°C, all others at 22°C. (C) Frequency sweep for

compounds 13 and 14 in the linear viscoelastic regime. Compound 13 strain amplitude: 0.05%; compound 14 strain amplitude: 0.5%. Solid and open symbols indicate the storage and loss modulus, respectively. (D) Frequency sweep for compound No. 8. Solid and open symbols indicate the storage and loss modulus, respectively.

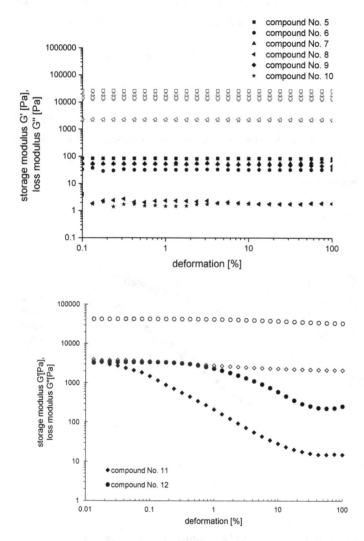

Figure 26 Amplitude sweep for the ionic liquids with compounds No. 6-12 (top) and for compounds 13 and 14 at a frequency of 1 Hz (bottom). Hollow and open symbols indicate the storage and loss modulus, respectively. $(Q^7)_5[\alpha\text{-SiW}_{11}O_{39}Fe(H_2O)]$ **(8)** was measured at T = 31°C, all others at 22°C.

Temperature dependence of the frequency sweeps was studied for compounds **6-14** and are exemplified using compound **8**, see Figure 27. To create a master curve, time-temperature superposition was carried out. With the temperature shift factor for the storage modulus G' and the loss modulus G'' taken to be unity, the vertical temperature shift factor a_T was determined at low frequency from the complex viscosity.

$$\eta_0 * : a_T = \left| \eta_0 * (T) \right| / \left| \eta_0 * (T_0) \right|_{,[70,71]}$$

T gives the temperature and T_0 gives the reference temperature. The absolute value of the complex viscosity was determined at 1 Hz as it is constant at the lowest frequencies studied. It was observed that the dominant loss moduli collapse on a master curve, suggesting Maxwell fluid behavior in the low frequency limit over the entire measurement range.

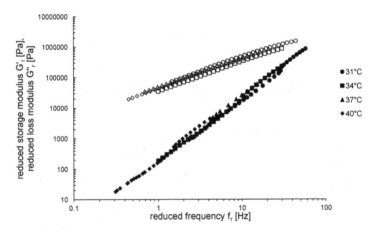

Figure 27 Master curve for $(Q^7)_5[\alpha\text{-}SiW_{11}O_{39}Fe(H_2O)]$ (**8**) with the storage modulus (solid symbols) and loss modulus (open symbols). The reference temperature is 34 °C.

The temperature dependence can be fitted with the Arrhenius function:

$$\ln a_T = \frac{E_a}{R}\left(\frac{1}{T} - \frac{1}{T_0}\right)$$

where E_a is the activation energy, R is the ideal gas constant and the temperature T is given in Kelvin. The reference temperature T_0 was chosen to be 31°C. The activation energy depends on the cation-anion interactions and can be interpreted as the average energy barrier that must be overcome for ions within the ionic liquid to move past each other.[67] The activation energies for compounds **6-14** are in the range of 77-109 kJ/mol, see Table 6 on the next page.

In summary, compounds **6-10+12** (Q^7-based) and **13-14** (Q^8-based) show qualitative differences in their rheological behavior. This is suggested to be due to the differences in cation alkyl chain length and in consequence, due to the cation-anion distances and interactions. For the Q^7-based POM-ILs **5-10,** viscoelastic behavior over a wide range of deformation was observed. In contrast, Q^8-based compounds **11-12** are less viscous and their internal structure is only stable at low deformations. At higher deformation rates, deviation from Maxwell fluid behavior, most likely associated with significant changes of the internal fluid structure, was observed.

Compound no.	Activation energy E_a [kJ/mol]
6	94±0.3
7	109±0.7
8	98±0.6
9	140±0.3
10	96±0.3
12	76.3±0.2
13	81±0.6
14	81±0.2

Table 6 The activation energy E_a of the compounds No. 6-14

3.1.4.5 Electromagnetic properties

EPR spectroscopy was used for investigation of the electromagnetic properties of the samples. All samples were dried as a thin film for two days under vacuum and solutions in toluene with a concentration of roughly $1 \cdot 10^{-3}$ mol/l prepared directly prior to use. All spectra were recorded at a temperature of T = 95 K.

{$SiW_{11}Cu$} as a $Cu-d^9$ system is EPR active with a spin state of 1/2 and shows a slightly anisotropic spectrum with a g-value of g($\perp$)= 2.35 and g(II) = 2.48. Hyperfine coupling with the Cu nucleus (spin I = 3/2) is seen and results in splitting of 4 lines in accordance to the selection rules. This is nicely seen for the parallel part of the g-tensor and to some extend also for the perpendicular component which normally shows line broadening to such extend that A($\perp$) is not resolved.

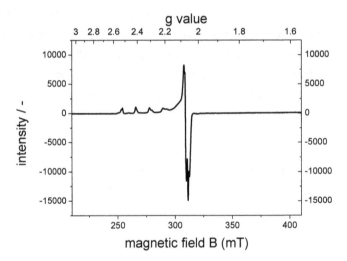

Figure 28: EPR-spectrum of $(Q^7)_6[\alpha\text{-}SiW_{11}O_{39}Cu(H_2O)]$ obtained in toluene (c = 1 mM) at a temperature of T = 95 K; radiation frequency f = 8943.7 MHz.

In addition superhyperfinecoupling is seen to some degree with further splitting of each peak into two signals. It may be possible that this interaction is due to the coupling between the electron spin with the heteroatom ^{29}Si (I = 1/2).

For {SiW$_{11}$Co} no EPR signal can be seen which comes as a surprise as Co(II) with a d^7 electron configuration should inherently be paramagnetic and thus give an EPR signal. However if oxidation of Co(II) to Co(III) has occurred, all electrons should be paired assuming a distorted octahedral geometry for the ligand field. For this reason it is believed that the oxidation state of cobalt in the synthesized compound is actually +III as has already been assumed from the melting behavior of this sample (see discussion of the melting ranges in chapter 3.1.4.3).

Iron was used in its oxidation state of +II and +III. Both weren't stored under inert atmosphere and within some days the {SiW$_{11}$Fe^{II}} adopted the same color as {SiW$_{11}$Fe^{III}} purely from visual observations and also with identical UV-VIS spectra. If oxidation of Co(II) to Co(III) (E° = +1.808 V) occurs as assumed, oxidation of Fe(II) to Fe(III) (E° = +0.771 V) is quite natural indeed (values for redox-potentials taken from "Chemie, C. E. Mortimer, U. Müller, Thieme, 2003, p. 676). A comparison of both EPR spectra confirms this with practically the same g-values for both samples.

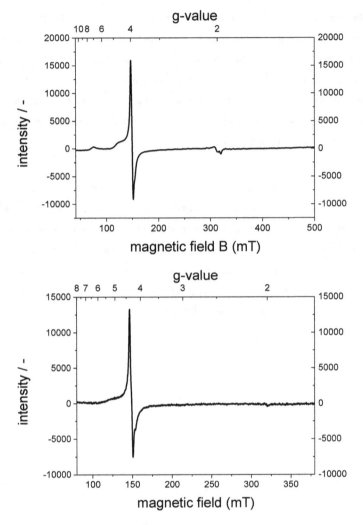

Figure 29: top: EPR-spectrum of $(Q^7)_6[\alpha\text{-SiW}_{11}O_{39}Fe(H_2O)]$ obtained in toluene ($c = 1\,mM$) at a temperature of $T = 95\,K$; radiation frequency $f = 8943.7\,MHz$; bottom: EPR-spectrum of $(Q^7)_5[\alpha\text{-SiW}_{11}O_{39}Fe(H_2O)]$ obtained in toluene ($c = 1\,mM$) at a temperature of $T = 95\,K$; radiation frequency $f = 8943.7\,MHz$.

The experimental g-values of $\{W_{11}Fe^{II}\}$ are found to be 2.024, 4.281 plus a weak signal at 8.48. This means that predominantly a spin $S = 3/2$ system is present, which for $Fe^{III}(d^5)$ may be the case for a octahedral deformed ligand field (C_{4v}), where the z-component is increased in energy. The character table for C_{4v} is given as follows together with the derived ligand field:

C_4v	E	$2C_4$	C_2	$2\sigma_v$	$2\sigma_d$		
A_1	1	1	1	1	1	z	x^2+y^2, z^2
A_2	1	1	1	-1	-1	R_z	
B_1	1	-1	1	1	-1		x^2-y^2
B_2	1	-1	1	-1	1		xy
E	2	0	-2	0	0	$(x, y)(R_x, R_y)$	(xz, yz)

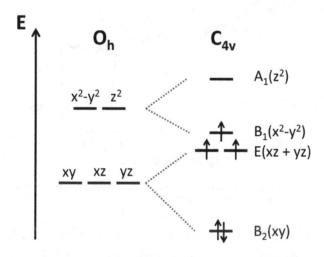

Figure 30: Splitting of d-orbital energies as derived for a C_{4v} symmetry with a strong ligand field in z-direction

The weak signal observed for g = 8.48 may be interpreted as a low lying excited state with S = 5/2.

With ^{57}Fe having a nuclear spin of I = 1/2, {SiW$_{11}$Fe} also shows hyperfinecoupling splitting the g = 2.024 signal into two as expected.

60

{SiW$_{11}$Ni} with Ni(II)-d^8 is EPR silent as would be expected from a diamagnetic compound taking the above derived energy diagram (see Figure 30) as basis.

{SiW11Cr} containing Cr(III)-d^3 gives a signal however a very poor one as is often the case for Cr-containing compounds. The g-values are found to be in the region of 2 and 4.5 allowing at least an unambiguous hint to a d^3 high-spin state.

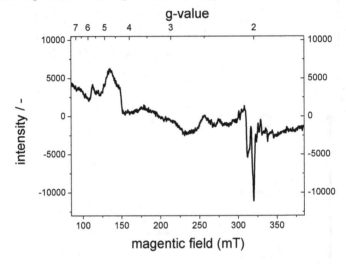

Figure 31: EPR-spectrum of (Q^7)$_5$[α-SiW$_{11}$O$_{39}$Cr(H$_2$O)] obtained in toluene (c = 1 mM) at a temperature of T = 95 K; radiation frequency f = 8943.7 MHz.

{SiW$_{11}$Mn} contains Mn(II)-d^5 which is isoelectronic to the Fe(III)-based analogue. The corresponding g-values are found to be in the same region with g = 2.03 and 4.17 also speaking for a spin S = 3/2 system with above shown energy diagram (see figure 32). Hyperfine coupling is seen with splitting into 6 lines as expected for ^{55}Mn (I = 5/2).

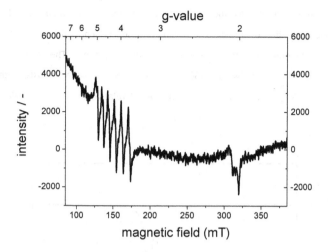

Figure 32 EPR-spectrum of $(Q^7)_6[\alpha\text{-SiW}_{11}O_{39}Mn(H_2O)]$ obtained in toluene (c = 1 mM) at a temperature of T = 95 K; radiation frequency f = 8943.7 MHz.

For Fe-containing samples Fe-57 Mössbauer spectrometry is an additional convenient analytic way to get insight into the electronic properties in this case first and foremost the oxidation state. Mössbauer spectrometry is the excitation of the nucleus with γ-radiation. The excitation energy is a function of electron density at the nuclear site or in other words Mössbauer spectrometry provides a means of measuring s-electron density at the nucleus as expressed in the formula for the isomer-shift δ:

$$\delta = const.\left(\frac{\Delta R}{R}\right)\left[\left|\Psi(0)\right|^2 - C\right]$$

Scheme 5: Formula for the isomer shift δ (mms^{-1}); ($\Delta R/R$): relative difference of nuclear radii between excited nucleus and nucleus in its ground state; $\left|\Psi(0)\right|^2$: charge density at the nucleus given by the square of the s-orbital wave function; C = constant term.

For ^{57}Fe the factor ($\Delta R/R$) bears a negative sign so that δ decreases with rising oxidation state.

^{57}Fe Mößbauer spectra were recorded by Dr. Jörg Sutter on a WissEl Mößbauer spectrometer (MRG-500) at 77 K in constant acceleration mode. ^{57}Co/Rh was used as the radiation source. WinNormos for Igor Pro software has been used for the quantitative evaluation of the spectral parameters (least-squares fitting to Lorentzian peaks). The minimum experimental line widths were 0.20 mms^{-1}. The temperature of the samples was controlled by an MBBC-HE0106 MÖSSBAUER He/N$_2$ cryostat within an accuracy of ±0.3 K. Isomer shifts were determined relative to α-iron at 298 K. The following spectrum was obtained:

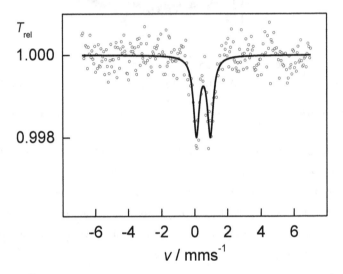

Figure 33 ^{57}Fe Mössbauer spectrum of (THA)$_5$[SiW$_{11}$O$_{39}$Fe], Type: doublet, δ = 0.53(1) mms^{-1}, ΔE$_Q$ = 0.84(1) mms^{-1}, Γ$_{FWHM}$ = 0.42(1) mms^{-1}.

Acquisition of a proper spectrum turned out to be difficult for the Fe-Keggin cluster as the eleven tungsten atoms (atomic number of 74!) per cluster effectively shield the iron nucleus from the gamma radiation, decreasing the effective cross section and increasing radiation loss due to *Compton scattering*. Therefore only a very thin film of the ionic liquid could be used, considerably decreasing the amount of active iron which resulted in a very weak signal. Yet it was possible to detect a response for {W$_{11}$Fe^{III}} when measured for five days. The value for the isomer shift was found to be δ = 0.53 mms^{-1}. In addition quadrupole splitting of ΔE$_Q$ = 0.84 mms^{-1} is observed which can be attributed to minor extend to the non-cubic ligand symmetry surrounding the Fe-centre. The main reason however should be found in an unsymmetrical electron distribution of the partially filled valance orbitals of the Mössbauer atom going hand in hand with the derived valance orbital scheme, where electron charge would be centered in the x-y-plane (see Figure 30).

3.1.5 Reactivity of the Si-templated Keggin clusters
3.1.5.1 Catalytic activity

Preliminary studies towards the catalytic activity of the Keggin based POM-ILs were carried out at room temperature in toluene/DMSO (6:1) by monitoring the light-induced degradation process of the commercial dye patent blue V (PBV) in the presence of the corresponding cluster. PBV was used in tenfold excess and the progress of decolorisation of the purple solution of PBV detected by measuring the absorbance of the prepared sample after 2 days under preclusion of light using UV-Vis spectroscopy. With $\{W_{11}Co\}$ a change of the spectrum was observed. Simultaneous UV-Vis measurements of a blank sample under identical reaction conditions (preclusion of sunlight, room temperature) without addition of any catalyst showed a negligible change of the spectrum.

The UV-Vis spectrum for the corresponding solution and the spectra of the blank solution without $\{W_{11}Co\}$ are shown in Figure 34.

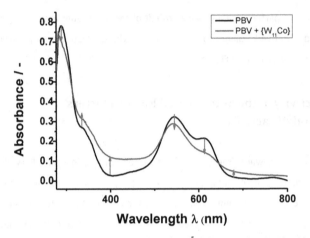

Figure 34: UV-Vis spectrum for blank PBV with c = $5 \cdot 10^{-5}$ M (black line) and with additional $\{W_{11}Co\}$ c = $1.55 \cdot 10^{-6}$ M (red line) after 2 d and under preclusion of light.

The study gives first insight into the potential of the catalytic activity. Further studies may investigate the photochemical degradation with light irradiation and time dependent experiments to elucidate the kinetics and gain a more thorough view on the catalytic behavior.

3.1.5.2 Reaction with CO

A position particularly interesting for chemical modification of the lacunary Keggin cluster is the sixth coordination site at which a water ligand is normally bound (see chapter 3.1.3.2). The water ligand can readily be substituted for other ligands like SCN⁻, CN⁻ or SO_3^{2-}.[59] Interestingly no studies towards the reactivity with carbon monoxide exist to the best of our knowledge.

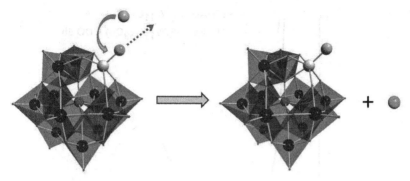

Figure 35 Ligand substitution reaction at the d-metal heteroelement (grey) of the cluster. The attacking ligand (orange) replaces an outgoing weaker ligand (green) and leaves the overall cluster structure intact.

In order to react the cluster with CO a dilute solution of compound **6**, **8** and **12** as well as a sample of the monolacunary Q^7{SiW$_{11}$} **5** were respectively prepared in toluene in a schlenk flask sealed via septum. CO-gas was bubbled in a constant weak stream through the solution via a syringe under atmospheric pressure and with magnetic stirring. After 5h reaction time the reaction mixture was immediately freed of solvent and dried in the same manner as the former ionic liquid.

All samples were again obtained as ionic liquids. For the turquoise Q^7{SiW$_{11}$Cu} a color change towards green, for the purple Q^7{SiW$_{11}$Mn} a color change towards red and for the lacunary Q^7{SiW$_{11}$} a color change from white to weakly brownish were observed. The yellow colored Q^7{SiW$_{11}$Fe} didn't show any visual changes. The change of color could also be evidenced by the compounds UV-Vis spectra (see Figure 36). It has to be noted that such reactivity was only observed in solution. If a solid compound e.g. the $(Q^6)_6[SiW_{11}O_{39}Cu(H_2O)]$ was reacted for the same time under a constant stream of CO gas no reaction took place (with unaffected UV-Vis and IR spectra).

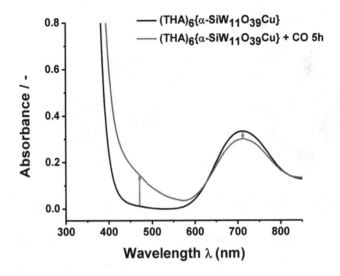

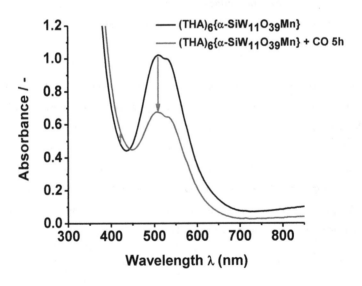

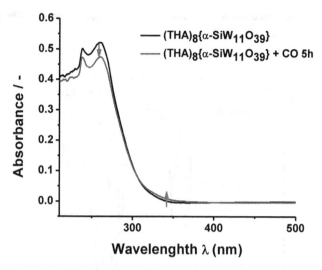

Figure 36 Comparison of the UV-Vis spectra of compound 5, 6 and 12 before (black line) and after (red line) reaction with CO gas.

As the spectra of Figure 36 show, the intensity of the former d → d excitation peak loses intensity, whereas the absorption in the LMCT region starts to increase, indicating a potential ligand exchange reaction with the water ligand being (partially) substituted. The fact that no additional d → d absorption is seen beside the decrease of the former absorption band however suggests that no stable coordination of the CO at the hetero d-metal occurs.

An interesting observation in this regard is that the "naked" $Q^7\{SiW_{11}\}$ also shows reactivity towards CO. Its reaction product exhibits an increased absorption intensity between ca. 300-400 nm and a decrease for the cluster centered LMCT peaks, which could suggest a partial reduction or even decomposition of the clusters overall structure. However no formation of precipitate was seen as was observed if the cluster was decomposed by addition of reducing agents like hydrazine. So it can be assumed, that the overall cluster structure is staying intact, which is also evidenced by the IR spectra (see below).

In order to gain more clarity if CO actually binds to the d-metal in the lacuna, a comparison of the ILs IR-spectrum before and after reaction with CO was obtained. The measurement was carried out on a Shimadzu FT-IR-8400S spectrometer in KBr pellet and transmission mode.

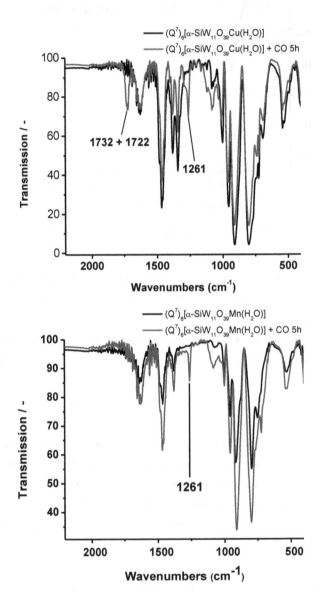

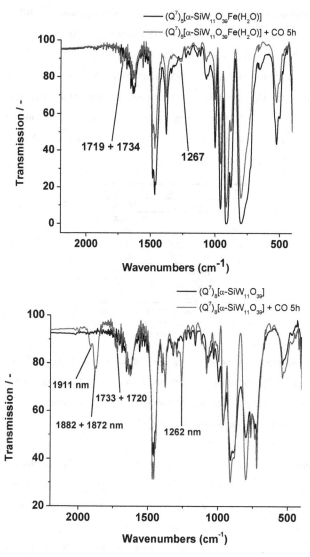

Figure 37 Comparison of the IR spectra of compound 5, 6, 8 and 12 before (black line) and after (red line) reaction with CO gas.

For all four compounds tested, the IR-spectra feature mainly identical absorption peaks before and after reaction with CO, which is a good indicator for the integrity of the cluster under these conditions. Its reaction with CO is indicated by three additional peaks located at ca. 1730, 1720 and 1260 nm. The Mn-substituted cluster $(Q^7)_6[\alpha$-

$SiW_{11}O_{39}Mn]$ interestingly only shows one additional absorption band at 1261 nm after the reaction with CO. For the underivatized $(Q^7)_8[\alpha\text{-}SiW_{11}O_{39}]$ additional absorptions occur at 1922, 1882 and 1872 nm after the reaction with CO.

When the electron density on a metal center bound to a CO-ligand increases, stronger π-back-bonding to the CO ligand is enabled which weakens the C-O bond by donation of more electron density into the formally empty carbonyl π^*-orbital. Therefore the M-CO bond strength is increased, making it more double-bond-like, and causes the resonance structure M=C=O to assume more importance.

d^x	compound	υ_{CO} (cm^{-1})
	free CO	2143
d^{10}	$[Ag(CO)]^+$	2204
d^{10}	$Ni(CO)_4$	2060
d^{10}	$[Co(CO)_4]^-$	1890
d^{10}	$[Fe(CO)_4]^{2-}$	1790
d^6	$[Mn(CO)_6]^+$	2090
d^6	$Cr(CO)_6$	2000
d^6	$[V(CO)_6]^-$	1860

Table 7 Overview of some homoleptic d-metal CO complexes with the wavenumbers for their carbonyl IR-vibration.

This effect is reflected in a weakened carbonyl IR-vibration e.g. the d^{10} complexes $[Fe(CO)_4]^{2-}$ and $[Ag(CO)]^+$ differ by over 400 cm^{-1} (see table 7). In addition a difference between equally charged complexes of different d-electron configuration is the consequence e.g. 60 cm^{-1} between $Ni(CO)_4$ and $Cr(CO)_6$.

Taking this well known relationship as basis it has to be assumed that the CO is not coordinating to the hetero d-metal as all additional peaks in the obtained spectra appear at practically the same wavenumber and do not show any dependency of the heterometals charge density. Reaction at a different position of the cluster, as was already assumed based on the missing shift of the d-d-absorption of the UV-Vis spectra, seems to take place. Complexes of tungsten with CO are not uncommon (one just has to think of the *Kubas*-complex) and in principle reaction of the CO with the addenda atoms W^{6+} could be possible in a reductive carbonylation reaction with $W(CO)_6$ as the product. The second IR-band appearing at around 1261 cm^{-1} could then be explained to stem from the oxidized reducing agent necessary for such type of reaction which would concretely be an oxidation product of the Q^7. However the IR-vibration of $W(CO)_6$ is located at $1998\ cm^{-1}$ and the observed peak at 1730 nm would actually speak for a strongly weakened CO-bond and indicate a bridging μ^2- or μ^3-CO complex.

For a complete structural analyses of the product and clarification of the fate of the CO a crystal structure analyses would be necessary which is nearly not possible to obtain for ionic liquids (as they are hardly forming crystals). A solution could be to charge balance the clusters with tetrabutylammonium which also enables phase transfer into the organic phase but yields solid products that can readily be crystallized by conventional solvent diffusion methods. Because the thesis at hand deals with ionic liquids this time consuming procedure was yet not carried out due to the restricted time schedule of the work. Mass spectrometry was utilized instead in order to obtain some more clarity of the reaction which the carbon monoxide could undergo with the investigated ionic liquids.

The used mass spectrometer has an ultra-high resolution (R(FWHM) ca. $4.1 \cdot 10^4$) with a mass accuracy error of 1 to 10 ppm depending on the quality of calibration. Detection was in negative-ion mode and the source voltage was 3.5 kV. The flow rates were 180 μl/hour. The drying gas (N_2), to aid solvent removal, was held at 180 °C with a flow rate of 4.0 l/min. The machine was calibrated prior to the experiment via direct infusion of the Agilent ESI-TOF low concentration tuning mixture, which provided an m/z range of singly charged peaks up to 2700 Da in both ion modes. Figure 38 shows the mass spectrum obtained from $(Q^7)_6[SiW_{11}O_{39}Cu(H_2O)]$.

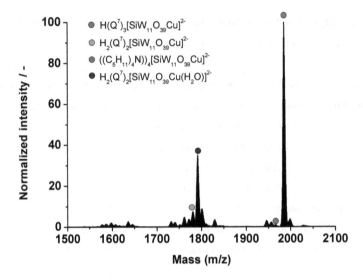

Figure 38 ESI-MS spectrum of the middle m/z region of $(Q^7)_6[SiW_{11}O_{39}Cu(H_2O)]$ taken in acetonitrile (concentration ca. 0.1 mM).

found m/z	calculated m/z	fragment
1780.0146	1780.0627	$H_2(Q^7)_2[SiW_{11}O_{39}Cu]^{2-}$
4382.9647	4382.9559	$H_2(Q^7)_3[SiW_{11}O_{39}Cu(H_2O)]^-$
1965.2617	1965.2773	$((C_5H_{11})_4N)_4[SiW_{11}O_{39}Cu]^{2-}$
1985.2169	1985.2962	$H(Q^7)_3[SiW_{11}O_{39}Cu]^{2-}$
1837.0326	1837.0364	$(C_3H_7)H_2(Q^7)_2[SiW_{11}O_{39}CuBr]^{2-}$
1790.4817	1790.5786	$H_2(Q^7)_2[SiW_{11}O_{39}Cu(H_2O)]^{2-}$

Table 8: ESI-MS peak list of assigned fragments in the middle m/z region with corresponding m/z values.

The mass spectrum after the reaction shows only very slight differences.

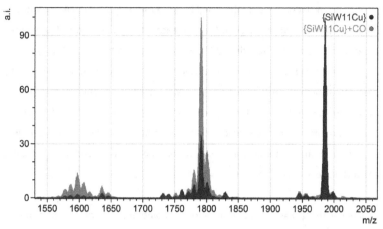

Figure 39: Comparison of the ESI-MS spectra of $(Q^7)_6[SiW_{11}O_{39}Cu(H_2O)]$ taken in acetonitrile before (blue) and after (orange) reaction with CO. Shown is only the m/z region from 1500 to 2100 where minor changes occurred.

Additionally ^{13}C-NMR spectra were comparatively recorded of the clusters before and after the reaction with CO which also did not show any obvious changes. No clear additional ^{13}C peak could be assigned. It seems possible that the degree the cluster is reacting is too low in both cases. So it will be necessary to ensure more conversion of the starting material by longer reaction times with the CO gas and by applying pressure in order to shift equilibria to the product side. Isolation of the CO-reaction product would be useful but is however nearly not possible from the ionic liquid medium.

3.1.5.3 Reaction with azide

A goal pursued by many inorganic chemists is the synthesis of high valent iron species with iron in its highest oxidation states of +IV, +V or even +VI. Such systems can function as model systems for biological processes where high valent iron species play a crucial role, or be useful in the activation of small molecules.[72]

Stabilizing the high oxidation state is a prerequisite for proving its existence and for making it chemically available. Many complexes in the current high valent iron research are based on organic ligands that are prone to oxidation or insertion reactions, quenching high oxidation states of the iron. Clusters as all inorganic materials could play a helpful role in this field of research. For this reason it was intended to first bind an azide ligand on the iron(III) substituted Keggin cluster and afterwards photolyze it to obtain the corresponding nitrido complex, which would correspond to an oxidation by two states yielding the iron(V)-nitrido complex. Making use of the ionic liquid medium is crucial as it excludes water as solvent which would compete with the azide for the free coordination position or quench higher iron oxidation states. Therefore great care was taken to fully exclude external water and all reaction steps were carried out using schlenk technique and dried solvents.

The reaction with N_3^- was carried out by stirring a solution of **8** in dry chloroform over an excess of NaN_3 for 4 h which resulted in a light color change from brownish yellow to an orange color (**8b**). After filtering off not reacted NaN_3 from the reaction mixture was one part of the solution evaporated to dryness. An IR-spectrum was obtained from this product in a KBr pellet prepared under glovebox atmosphere (IR-1). The other (still dissolved) part was transferred into a quartz tube and irradiated for 24 h using a 150 W medium-pressure mercury light source (Heraeus TQ-150, UV Consulting Peschl λ_{max} = 365 nm equipped with a pyrex deep-UV cutoff filter λ_{cutoff} = 320 nm) which resulted in a color change from orange/yellow to deep blue (**8c**). An IR-spectrum was prepared in the same manner (IR-2). The blue colored compound is shave-stable under exclusion from air and starts to precipitate slowly over several days. Contact with air leads to a very fast reaction yielding again a compound of yellow color within around 5 seconds.

IR-1 shows an additional strong peak at a wavenumber of 2060 cm^{-1} compared to the precursor **8**, which is a good indicator that the azide ligand could have coordinated to the iron centre, particularly with respect to the additional color change of the reaction solution. After the photolyzation this peak has completely vanished in IR-2 and no peak is seen in the region from 2300-1800 cm^{-1}.

The color change of all steps was monitored by UV-Vis spectroscopy. Starting compound **8** shows a LMCT transition at 285 nm and a d → d transition falling together with the boarders of this peak (see Figure 40 left spectrum). After the reaction with NaN_3 a bathochromic shift has occurred and the d → d transition peak is now clearly separated at a wavelength of 405 nm. Photolysis of the reaction mixture leads to a further strong batochromic shift. Two absorption peaks are found at 617 and 889 nm (see Figure 40 right part).

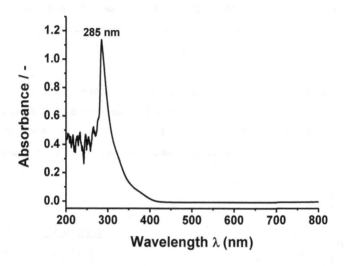

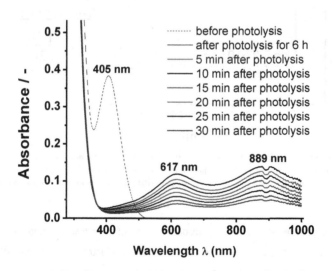

Figure 40 top: UV-Vis spectrum of the precursor **8** in toluene (c = 6.2 10^{-6} M). bottom: dotted line showing the UV-Vis spectrum of **8b**; black line showing the UV-Vis spectrum of **8c**, colored lines showing the time dependent decomposition of **8c** on air monitored via its UV-Vis spectrum.

The decomposition of **8c** on air could also be monitored by UV-Vis spectroscopy. A quartz glass cuvette was filled under N_2-inert atmosphere with a solution of the photolyzed compound and sealed with a teflon stopper. Placed outside inert atmosphere enough air permeates the sealed cuvette to monitor the decomposition on a minute time scale.

The decomposition follows a zero-order kinetics with respect to the substrate **8c** (see Figure 41 on following page) with the reaction rate being determined by the amount of air permeating the cuvette. The rate constant k was calculated to be $3.92 \cdot 10^{-5}$ s^{-1}. A possible field for application of this highly sensitive reaction could be found in sensor systems.

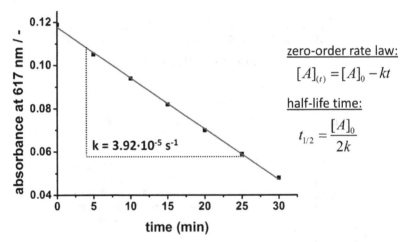

zero-order rate law:

$$[A]_{(t)} = [A]_0 - kt$$

half-life time:

$$t_{1/2} = \frac{[A]_0}{2k}$$

k = 3.92·10⁻⁵ s⁻¹

Figure 41: Plot of the absorbance values at 617 nm from the time dependent decomposition of **8c** on air in the time period of 0 min - 30 min (black squares) together with the linear fit (red line). It shows zero-order kinetics and the associated equations are given on the right side. The rate constant k was calculated to be $3.92 \cdot 10^{-5}$ s^{-1}.

In order to determine whether the oxidation state of the cluster bound iron had changed after photolysis of **8b**, a second Mössbauer spectrum was comparatively measured after photolysis. Both spectra are given in Figure 42 (see next page).

Again only a very thin film of the sample **8c** could be applied in the measurement due to shielding from the tungsten atoms, as has already been described in chapter 3.1.4.5. Yet it was possible to clearly acquire a response. A comparison with the starting material **8** shows a shift of the isomer shift δ of +0.52 mms^{-1} from 0.53 mms^{-1} to 1.05 mms^{-1}. As ^{57}Fe bears a negative sign for (δR/R) a positive isomer shift implies a decrease of the electron density at the nucleus, which means that the oxidation state of the iron centre is decreased in compound **8c** (containing Fe in its oxidation state of +III) compared to compound **8**. Therefore it is proposed that a photochemically controlled reduction process takes place with the azide(ligand) being oxidized. The strong color change from orange to dark blue can be explained by formation of a so called "heteropoly blue". It is known that numerous heteropoly anions can be reduced by one or more electrons, which are delocalized, according to various time scales, over certain atoms or regions of the cluster. The reduction products, typically retain the general structures of their oxidized parents and have a dark blue color.[73, 74] In addition to the higher isomer shift a strongly increased quadrupole splitting occurs for the photolyzed product **8c**. It can be attributed to a decrease of ligand symmetry surrounding the Fe-centre with azide or a vacancy at the sixth coordination position for compound **8c** compared to the iron being coordinated to six oxygens in compound **8**. Because of the strong splitting it is assumed that the additional electron(s) add to an unsymmetrical electron distribution of the partially filled valance orbitals of the iron center.

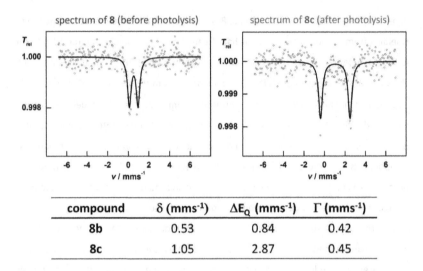

compound	δ (mms⁻¹)	ΔE_Q (mms⁻¹)	Γ (mms⁻¹)
8b	0.53	0.84	0.42
8c	1.05	2.87	0.45

Figure 42 Mössbauer spectra of compound 8b (left) and 8c (right) obtained at 77 K in constant acceleration mode with ^{57}Co/Rh as the radiation source. The isomer shift δ, the quadrupole splitting ΔE_Q and the line width Γ are tabulated. Isomer shifts were determined relative to α–iron at 298 K.

Due to the coordination environment and electron properties POMs provide to the d-heterometal with the substituted metal atom coordinated about its equator into an electron-conducting structure do POMs show strong analogy to metalloporphyrins.[75] For Fe(II) S=2 heme-complexes values of δ = 0.85-1.0 mms⁻¹ and ΔE_Q = 1.5-3.0 mms⁻¹ are common.[76] Therefore it is assumed that the physical oxidation state of the iron center in compound 8c is +II and one reduction electron of the heteropolyblue is mainly centered at the iron. For more details, as on how many electrons are transferred to the cluster and further information about their position, additional analytic investigations comprising EPR, SQUID and ^{183}W-NMR will be necessary.

3.2. Variation of the templating heteroelement - the B and P containing Keggin clusters

Varying the templating heteroelement of the Keggin cluster was done in order to change the clusters charge without significantly changing the anions surface charge density or its radius.

No.	cluster	heteroelement	charge of cluster	ref. for synthesis
22/23	{BW$_{11}$}	B^{3+}	-9	[77]
1-14 + 20/21	{SiW$_{11}$}	Si^{4+}	-8	[56]
24/25	{PW$_{11}$}	P^{5+}	-7	[78]

Table 9: Overview of the differently templated Keggin clusters used in this work, with listing of the heteroelements and overall cluster charge as well as the reference for synthesis.

For the synthesis of {PW$_{11}$Cu} **(24)** an alternative direct synthetic route was tested. Instead of synthesizing the monolacunary {PW$_{11}$} and isolating it in a first step, the {PW$_{11}$Cu} was formed in situ according to the literature reported procedure [78] and directly phase-transferred from the reaction mixture with Q^7-Br dissolved in toluene.

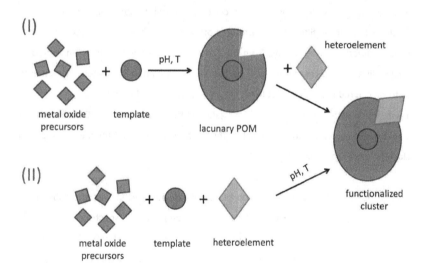

Figure 43 The two different synthetic routes compared. (I) The two step reaction with isolation of the lacunary POM prior to functionalization with the heteroelement (orange rhomb). (II) One step formation of the functionalized POM directly out of the reaction solution containing the metal oxide precursor (blue squares), the templating ion (purple circle) and the heteroelement (orange rhomb).

The elemental analysis of the product obtained after removal of solvent under vacuum and several times lyophilizing shows more impurity compared to the two step procedure.

EA of **24** (calculated values in brackets): C: 33.09 (34.94), H: 5.98 (6.32), N: 1.23 (1.46).

It is however a relatively fast way of synthesizing the POM-IL and may suffice for preliminary studies e.g. concerning catalytic activity or chemical behavior of a POM-IL. If absolutely clean products are needed (e.g. for rheological measurements, determination of chemophysical properties or electrochemical investigations) crystallization (and possibly recrystallization) of the functionalized cluster is necessary prior to its phase transfer. Particular attention has then to be paid that the amount of crystal bound water is exactly determined e.g. by *Karl-Fischer* titration or TGA, which makes the alleged simpler one step synthesis rather intricate too.

For the synthesis of the mono-lacunary $K_7[PW_{11}O_{39}]\cdot nH_2O$ an aqueous solution of dodecatungstophosphoric acid has to be brought to a pH of 5 with $KHCO_3$, which already causes the formation of the cluster at room temperature. Insoluble side products have to be filtered or centrifuged off before the $K_7PW_{11}O_{39}\cdot nH_2O$ can directly be crystallized as white crystals from the concentrated reaction mixture. An additional recrystallization is recommended.[79] During the reaction no metal spatula shall be used as these can reduce the product (which is seen at hand of a blue coloring).
The reaction with a 3d-heterometal for ligation in the lacuna and the POMs final phase transfer into toluene works identical to the procedure described for the Si-templated clusters (see chapter 3.1.1 & 3.1.2.).

Insight into the electronic interaction of the phosphate oxygen with the substituent metal in the lacuna can be gained at hand of the compounds IR spectra as *Weakley* [80] and *Rocchicciolo-Deltcheff* [81] have shown. It is literature known that the P-O stretching vibration at 1080 cm^{-1} in the spectrum of $[PW_{12}O_{40}]^{3-}$ is split by 45 cm^{-1} in $[PW_{11}O_{49}]^{7-}$.[80] Examination of the obtained vibrational spectra of (**24**) and (**25**) shows a similar splitting of 75 cm^{-1} for the $(Q^7)_5[PW_{11}O_{39}Cu(H_2O)]$ whereas the spectrum of the $(Q^7)_4[PW_{11}O_{39}Fe(H_2O)]$ more closely resembles this of the $PW_{12}O_{40}^{3-}$ with one distinct P-O stretching vibration at 1069 cm^{-1}.

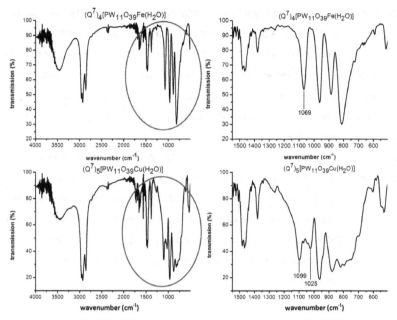

Figure 44 Comparison of the IR spectra of (**24**) and (**25**). At the right side the IR-region containing prominent cluster vibrations (encircled in red in the full spectrum at the left side) is enlarged.

The authors argued that the magnitude of this splitting can be taken as a measure of the lack of interaction of the phosphate oxygen with the heterometal that depends on how good the heterometal fits into the lacuna. So it can be assumed that the Fe(III) in $(Q^7)_4[PW_{11}O_{39}Fe(H_2O)]$ shows better interaction with the phosphate oxygen than the Cu(II) in $(Q^7)_5[PW_{11}O_{39}Cu(H_2O)]$ which also seems reasonable with respect to the respective ionic radii (65 pm for Fe^{III} vs. 73 pm for Cu^{II})[82]. Concerning further studies of the chemical behavior of these POM-ILs this finding may play a helpful role in understanding their reactivity.

For synthesis of the POM-ILs containing bor-templated clusters the one step synthetic route (II) was chosen because the monolacunary {BW_{11}} is reported to be obtained only in poor yields and as an unstable product.[77] So the complete cluster was formed in situ by self-assembly of the WO_4^{2-}; BO_3^{3-} and Cu^{2+}/Fe^{3+} building blocks. It is well known that several factors such as the pH value, type of initial reactants and temperature have crucial effect on the final products in the synthesis of POMs. In this case the pH is the most important factor and should be carefully controlled at about 7.0, because the monolacunary $[BW_{11}O_{39}]^{9-}$ is only stable at a narrow pH range (pH 7–8). When the pH is lower than 6, it transforms into $[BW_{13}O_{46}H_3]^{8-}$ at room temperature or the saturated $[\alpha\text{-}BW_{12}O_{40}]^{5-}$ while increasing the temperature. When the pH is larger than 8, the polyanion may be destroyed only leaving borate and tungstate in solution.[83] There is also evidence that $[BW_{11}O_{39}]^{9-}$ is not existing in solution due to its high negative charge and is only stable as $[BW_{11}O_{39}H]^{8-}$.[84]

So the frail synthesis gives rise to expect a considerable amount of side products from the reaction why the functionalized clusters were crystallized from the solution and not directly phase transferred. For detailed synthetic instructions please see the experimental section as well as the adopted procedures from literature.[77, 85, 86] The bor templated POM-ILs could not be obtained in satisfying purity so that a comparison with the other ILs synthesized in this work concerning their physical properties isn't reasonable at this point.

3.3. Variation of the cation - phosphonium and imidazolium POM-ILs

Tetraalkylphosphonium polyoxometalate-based POM-ILs were first reported in 2007 as a different approach towards true POM-ILs. Here, the choice of organic cation was rather straight-forward as tetraalkylphosphonium salts of POMs are renowned for their facile preparation as well as their high thermal stability.[67, 87] The authors screened a range of imidazolium and phosphonium cations, several of which only led to high melting-point (> 200 °C) solids which were not further investigated. However, the combination of trihexyltetradecylphosphonium, $(C_{14}H_{29})(C_6H_{13})_3P^+$ cations with Keggin $[PW_{12}O_{40}]^{3-}$ and Lindqvist $[W_6O_{19}]^{2-}$ anions gave true POM-ILs with melting points of 65 °C (Keggin) and -48 °C (Lindqvist), respectively. Thermal stability measurements using TGA-DSC analysis showed that no weight loss is observed up to ca. 480 °C; long-term heating experiments further showed that **POM-IL-2** (see. chapter 1.4 for details) stays intact for prolonged periods at temperatures up to 360 °C, so that a drastically higher thermal stability is observed for **POM-IL-2** compared with **POM-IL-1**.

A high thermal stability is favorable or even necessary for many catalytic reactions. Therefore it was investigated if the combination of trihexyltetradecylphosphonium ions would also improve the thermal properties of the POM-ILs of this work and if the charge balancing with the applied Keggin ions would yield a ionic liquid at all. Therefore two additional reactions were carried out in which trihexyltetradecylphosphonium bromide $((C_{14}H_{29})(C_6H_{13})_3P)Br$ was used instead of Q^7-Br for the phase transfer. {SiW$_{11}$Cu} and {SiW$_{11}$Fe} were chosen as approved and reliable model clusters.

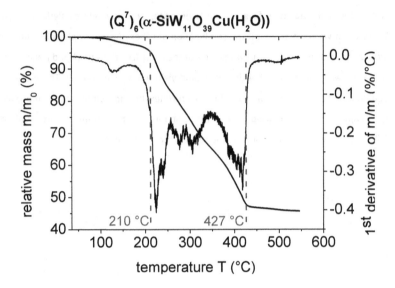

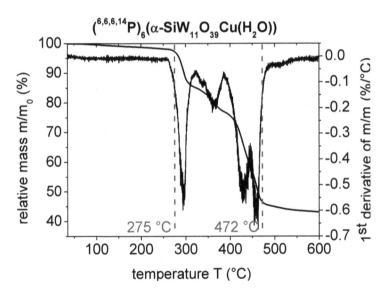

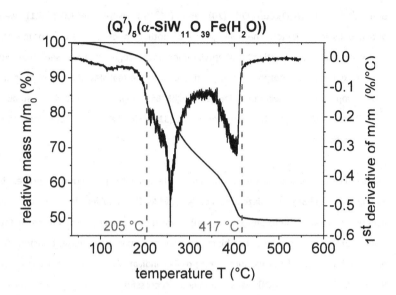

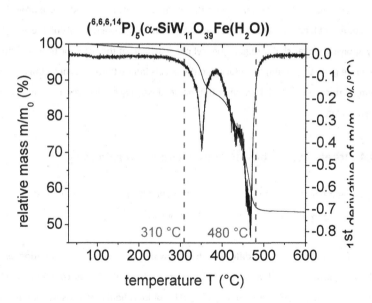

Figure 45 Comparison of the TGA-graphs for the Q[7]-based vs. the [6,6,6,14]P-based POM-ILs of [α-SiW$_{11}$O$_{39}$Cu(H$_2$O)]$^{6-}$ and [α-SiW$_{11}$O$_{39}$Fe(H$_2$O)]$^{5-}$ respectively. The relative weight loss with rising temperature (blue line) and its derivative (black line) are shown. Onset and end of the cations decomposition are marked with a dashed line (red).

Both $(^{6,6,6,14}P)_6[SiW_{11}O_{39}Cu(H_2O)]$ **(20)** and $(^{6,6,6,14}P)_5[SiW_{11}O_{39}Fe(H_2O)]$ **(21)** were obtained as ionic liquids. The TGA graphs of both compounds were recorded and compared with the respective tetraheptylammonium analogue. It can be seen that with the trihexyltetradecylphosphonium ions an improved thermal stability is gained. So the onset temperature for decomposition lies roughly 65 °C higher for **20** (275 °C instead of 210 °C) and even roughly 105 °C higher for **21** (310 °C instead of 205 °C) compared to the Q^7-based equivalents.

As a third cation species the imidazolium ion was tested in the phase transfer process by adding 1-methyl-3-octylimidazoliumchloride (OMIM-Cl) dissolved in toluene to an aqueous solution of the $[\alpha\text{-SiW}_{11}O_{39}Cr(H_2O)]^{5-}$ which caused a green solid to precipitate when both solutions where combined instead of initiating a phase transfer. The elemental analyses of the vacuum dried product indicates that the target compound $(OMIM)_5[\alpha\text{-SiW}_{11}O_{39}Cr(H_2O)]$ was obtained successfully. Therefore no further investigations where undertaken with the product and it is intended to use alkyl chains of a different length on the imidazolium ion in order to induce more steric pressure on it by applying longer chains (e.g. a dodecyl chain instead of the octyl) or to use a shorter chain (e.g. a ethyl chain) to reduce van der Waals interactions and thus obtain a room temperature ionic liquid. Each of both modifications may be the right path and have to be elucidated.

3.4. Other clusters beside monolacunary Keggin-ions

Applying other clusters beside the well established Keggin polyanion in POM-ILs was of interest mainly due to the following three reasons:

1) Different clusters carry different charges which means that the variation of the cluster anion goes hand in hand with a variation of the number of counterions, which plays a crucial role for the rheological behavior of ionic liquids (an observation that has once more been demonstrated with the results of this work).

2) A further important factor for a ionic liquids behavior is the degree of electrostatic interaction between its cations and anions, which is mainly determined by weak intermolecular interactions like van der Waals forces, Coulomb interactions or hydrogen

bonds. Therefore the question arose if a cluster of double the size as the Kegginion would also be obtained as an ionic liquid if paired with e.g. the tetraheptylammonium cation.

3) The successful reaction of the Keggin based POM-ILs with CO stimulated to search for systems which would offer better options for catalytic application. Even if it is currently not clear how the cluster actually reacts with CO it did seem reasonable that the 3d-transition metal is a likely binding site. The Keggin-cluster binds to the remaining five coordination positions leaving little scope for catalytic reactions. In order to react a substrate with the CO a further binding side in its proximity would be needed. For this reason it was tried to imply the dilacunary $\{SiW_{10}\}$ in a similar POM-IL.

3.4.1. Ionic liquids based on Dawson Clusters

As Keggin Clusters were historically named in honor of *J. F. Keggin* who was the first scientist solving the compounds structure is it the same story with Dawson type clusters who are named in honor of *B. Dawson* who solved the α and β-isomer structure of $K_6[P_2W_{18}O_{62}]$ $14H_2O$ in 1953 by X-ray crystallographic determination.[88]

The Dawson anion consists of two PO_4^{3-}-units encapsulated by a $[W_{18}O_{54}]$ cage, which itself is built up from two α-XW_9O_{34} half units joined by six common oxygen atoms, lying on a plane of symmetry. The whole anion belongs to D_{3h} point group and incorporates two types of tungsten atoms, 6 "polar" at the caps and 12 "equatorial" ones in between (see also Figure 46 left structure). The β–anion derives from the α isomer by a formal rotation by $\pi/3$ of a polar (cap) W_3O_{13} group: the symmetry is then lowered to C_{3v}. The formal rotation by $\pi/3$ of the second polar W_3O_{13} group restores the plane of symmetry and the point group D_{3h} for the γ isomer. In all these anions the hexagonal belts of both XW_9, moieties are symmetry related through the equatorial horizontal plane and their twelve tungsten atoms appear eclipsed along the direction of the C_3 axis.

Substitution of one polyhedral with a d-element, as was done with the Keggin-clusters in the previous section, is also possible for the Dawson-type systems with several examples existing in literature.[89-91]

88

For this the parent ion $[\alpha_2\text{-}P_2W_{18}O_{62}]^{6-}$ has to be converted into the defect complex $[\alpha_2\text{-}P_2W_{17}O_{61}]^{10-}$ which afterwards can be substituted with an additional element and extracted into organic media using Q^7-Br following an analogous reaction procedure as for the Keggin-POMs (for details see experimental section).

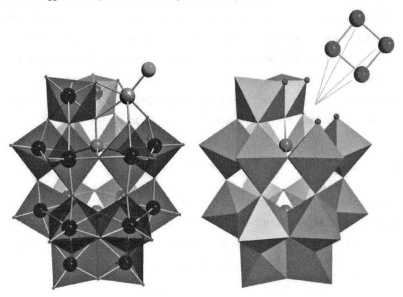

Figure 46 left: Ball and stick representation of the substituted Dawson cluster $[\alpha_2\text{-}P_2W_{17}O_{61}ML]^{10-}$; heteroatom phosphor (orange), addenda atoms tungsten (blue), oxygen (red), d-element M (grey), additional ligand L e.g. water (green). right: Polyhedral representation with $[WO_6]$-polyhedra (blue), heteroatom phosphor (orange), oxygen (red) and highlighting of the square lacunary site with its four main oxo-ligands.

In this work Co(II) and Fe(III) have been incorporated successfully into the Dawson cluster with consecutive phase transfer into toluene using Q^7-Br. Both products $(Q^7)_8[\alpha_2\text{-}P_2W_{17}O_{61}Co(H_2O)]$ and $(Q^7)_7[\alpha_2\text{-}P_2W_{17}O_{61}Fe(H_2O)]$ show clearly higher viscosity compared to the synthesized Keggin ILs and are obtained as gummy solids at room temperature. This finding may be rationalized at hand of the size and charge of the Dawson cluster. It has a prolate ellipsoidal shape (so called prolate spheroid) in which the polar axis is greater than the equatorial diameter. Its dimensions are ca. 12.2 Å x 10.3 Å x 10.3 Å corresponding to a volume of ca. 680 Å^3. The Keggin anion has dimensions of ca. 8.2 Å x 10.3 Å x 10.3 Å and thus a volume of ca. 455 Å^3. In the following Table 10 the average volume charge densities are compared calculated according to $\langle \rho_q \rangle = \frac{Q}{V}$. with Q = charge of the cluster anion and V being the volume of the cluster. Additionally the surface area A is calculated which is given for the biaxial prolate spheroid - with the longer polar axis denoted as a and the two shorter equatorial ones as c - as:

$$A = 2\pi c^2 \left(1 + \frac{a}{c} \frac{\sin^{-1}\epsilon}{\epsilon} \right) \text{ with } \epsilon = \sqrt{1 - \left(\frac{c}{a}\right)^2} \text{ and a > b = c}$$

cluster	type	3d-heterometal	cluster charge	ρ_q (C/Å^3)	A (nm^2)
$(Q^7)_6[\alpha\text{-SiW}_{11}O_{39}Co(H_2O)]$	Keggin	Co(II)	+6	$2.1 \cdot 10^{-21}$	330
$(Q^7)_8[\alpha_2\text{-P}_2W_{17}O_{61}Co(H_2O)]$	Dawson	Co(II)	+8	$1.9 \cdot 10^{-21}$	485
$(Q^7)_5[\alpha\text{-SiW}_{11}O_{39}Fe(H_2O)]$	Keggin	Fe(III)	+5	$1.8 \cdot 10^{-21}$	330
$(Q^7)_7[\alpha_2\text{-P}_2W_{17}O_{61}Fe(H_2O)]$	Dawson	Fe(III)	+7	$1.6 \cdot 10^{-21}$	485

Table 10 Average volume charge density ρ_q and surface area A compared between Keggin and Dawson clusters derivatized with Co(II) and Fe(III).

It is seen that the Dawson-type clusters feature average volume charge densities of around 10% lower than the respective Keggin-anions which would lead one to expect their respective POM-ILs to have the lower viscosity compared to the Keggin analogues. However does the Dawson structure have nearly 150% the size of the Keggin-ion which promotes stronger Van der Waals interactions for two reasons. First is the polarizability eased, leading to stronger London dispersion forces and better Debye interaction. And secondly can the sterically demanding cations better converge the cluster-anion which increases their coulomb attraction and additionally the Van der Waals interaction that is depended on the distance with approximately the power of six. The comparison at hand shows strikingly that the viscosity and melting points of ionic liquids are hardly predictable due to these opposing effects and that a cluster with less charge density may in the end yield a solid at room temperature, whereas a comparable cluster with less charge density yet of smaller size may yield a room temperature ionic liquid.

3.4.2. Ionic liquids based on the dilacunary {SiW$_{10}$}

All clusters discussed so far featured one lacunary position to bind additional d-metals, to which the cluster behaves as pentadentate ligand. This leaves only one single coordination position free for performing chemical reactions. For many reactions it would however be beneficial or even necessary to provide more than one binding position at which the reagent and substrate could be brought into close proximity for a reaction, e.g. CO-insertions or cross-coupling reactions. Aiming at improved possibilities for cluster based catalytic reactions in the ionic liquid medium it was indented to synthesize the dilacunary {γ-SiW$_{10}$}, functionalize it with a d-heterometal and obtain it as a POM-IL. The {γ-SiW$_{10}$} is synthesized by increasing the pH of an aqueous solution of the monolacunary K$_8$[β-SiW$_{11}$O$_{39}$]·14H$_2$O to pH = 9.1 with K$_2$CO$_3$ and precipitate the product after exactly 16 minutes.[56] This procedure leads to the formation of one additional void adjacent to the lacunary position of the starting material creating a rectangular lacuna with four main oxo-ligands at the edges (see Figure 47 right part). The lacuna can accommodate two heterometals. For experimental purposes Cu(II) was applied and 2.1 eq. of CuCl$_2$·2H$_2$O added following a similar reaction procedure as for the {SiW$_{11}$} clusters. For details see the experimental part.

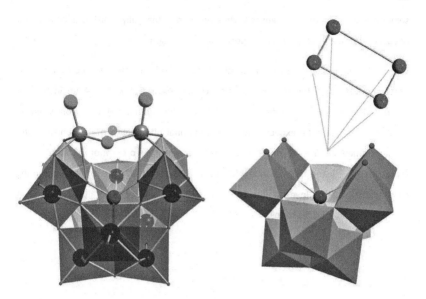

Figure 47 left: The disubstituted dilacunary Keggin cluster $[SiW_{10}O_{36}M_2]^x$. (left) Ball and stick representation with heteroatom silizium (pink), addenda atoms tungsten (blue), oxygen (red), d-element M (grey), additional ligands e.g. water or Cl^-/Br^- can be bound as terminal or bridging μ_2-ligands (green). right: Polyhedral representation with $[WO_6]$-polyhedra (blue), heteroatom silizium (pink), oxygen (red) and highlighting of the rectangular lacunary site with the four main oxo-ligands.

The $[SiW_{10}O_{36}Cu_2]^{4-}$ cluster was phase transferred into toluene with Q^7-Br and obtained as a room temperature ionic liquid (**27**). Each cupper metal can bind directly to three oxo-ligands of the cluster, namely two main oxo-ligands located on the short side of the rectangular lacuna and one slightly remote oxo-ligand of the templating phosphate. As additional ligands bound to the Cu(II) centers water as well as Cl^- (stemming from the $CuCl_2 \cdot 2H_2O$) or Br^- (stemming from the Q^7-Br) have to be considered and both terminal and bridging μ_2-positions are possible. The single negatively charged halogen ligands play a critical role in this respect as they alter the overall cluster charge and thus the number of charge balancing Q^7 counterions. The elemental analysis of the thoroughly vacuum dried compound **27** shows that halogen ligands may be present: EA in wt. - % (calculated values in brackets): C 35.02 (31.66), H 6.54 (5.79), N 1.41 (1.32). The obtained values speak for an average of five instead of four Q^7 cations which would mean that one additional halogen is bound on an average. Addition of $AgNO_3$ to a solution of **27** in acetonitrile causes a white precipitate that fully vanishes upon addition of concentrated ammonia which supports this assumption.

A slight batochromic shift of the absorption maximum in the UV-Vis spectrum of **27** is noted, compared to the monosubstituted cluster {SiW$_{11}$Cu} with 724 nm instead of 710 nm. The extinction coefficient of **27** lies higher with 27 M^{-1}cm^{-1} compared to 18 M^{-1}cm^{-1} as would be expected with an additional photoactive metal centre in the cluster framework. This finding supports the successful incorporation of a second d-metal in the cluster framework which turns compound **27** into a very promising candidate for further investigation concerning its applicability as catalyst or its use in photochemical based reactions.

4. Summary and Perspective

The focus of this work was to synthesize and characterize new POM based ionic liquids with thorough investigation of their physiochemical properties in order to create a basis for further studies of this relatively new compound class. Polyoxometalates being purely inorganic compounds could be of high interest for industrial use due to their stability and promising examples in catalysis and materials science. Transfer into organic reaction media is however a prerequisite for many applications. Typically, POMs form salts with high lattice energies and often low solubility in many common solvents. Converting POMs into ionic liquids is therefore an ideal route to deliver them into otherwise non-accessible reaction media. Access to pure POM-ILs is particularly attractive as they can combine high cluster loadings with unique chemical and rheological properties which open the door to an enhanced field of application. Today, it is not only possible to perform a new type of chemistry with the clusters but also to dissolve them in organic solvents. Therefore some of the main recent advances in the synthesis and application of true POM-ILs were summarized, with focus on catalysis and organic synthesis. A selection of three promising fields of current state of the art research was exemplarily given to highlight the impact of POM-IL science.

In a systematic study the influence of the POM-ILs different components was researched. Adjustment parameters for the tuning of POM-ILs are the organic cation and the cluster anion that can be modified with respect to size and shape or in other words the type of cluster as well as its elements like the type of addenda atom or the templating heteroelement (see Figure 48 on the next page).

94

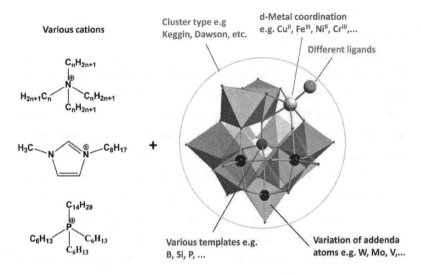

Figure 48 The various "screws" of a POM-IL which can be modified, like the type of cation (left structures) used for charge balancing the cluster (right structure shown in polyhedral representation with the front triad transparent).

With the Keggin-ion $[SiW_{11}O_{39}M(H_2O)]^{n-}$ as model system it was shown that the type of transition-metal M coordinated in the lacuna of the cluster does not have a significant influence on the POM-ILs rheological properties, its melting point, viscosity or the thermal stability, which is particularly helpful for designing catalytically active POM-ILs as the lacuna is a crucial part to perform coordination chemistry with the clusters. Based on detailed UV-Vis studies and a comparison with the corresponding hexaquaions known from aqueous solutions it could be demonstrated that the d-element heterometal M retains an octahedral coordination environment if the cluster is transferred into organic media.

The reaction of the Keggin based POM-ILs with CO highlighted a potential field of application. First steps for elucidating the mechanism were undertaken. Based on the UV-Vis spectra it has to be assumed that the cluster is not coordinating the CO on the hetero d-metal located in the lacuna but instead possibly reacting with its W-addenda atoms. Future synthetic effort may thus be directed to the modification of the cluster as to enhance the reactivity of the 3d-heterometal position. Applying tetrabutylammonium instead of tetraheptylammonium counterions could be used to obtain solid products which can then be characterized at hand of their crystal structure in order to gain clarity of the fate of the CO molecule. A further useful tool for elucidation of the reactive position in this reaction could be [183]W-NMR.

A similar approach may be useful for the photochemically guided reaction of the $(Q^7)_5[\alpha\text{-SiW}_{11}O_{39}Fe^{III}(H_2O)]$ with azide, which could be demonstrated and proved using UV-Vis and IR spectroscopy. It was shown via Mössbauer-spectroscopy that the photolyzation process leads to a photochemical reduction of the cluster bound Fe(III) to a heteropolyblue cluster containing iron in a physical oxidation state of +II.

Additional charge variation of the cluster by variation of the templating heteroelement in the systematic row of undecatungstoborate, undecatungstosilicate and undecatungstophosphate with Q^7-counter ions and bivalent Cu^{II} and trivalent Fe^{III} bound in the lacuna showed that the Keggin clusters can be obtained as an ionic liquid over a wide range of cluster charge if paired with the correct cation. Q^7 yielded ionic liquid products from the four times negative $[\alpha\text{-PW}_{11}O_{39}Fe(H_2O)]^{4-}$ up to the seven times negative $[\alpha\text{-BW}_{11}O_{39}Cu(H_2O)]^{7-}$. By applying Dawson type clusters another verification could be established that the cluster charge or rather its charge density is not playing the crucial role for the products physical properties rather than the volume and surface size which influences the degree of van der Waals interaction between cluster and organocation.

The final adjustment parameter in POM-IL composition is the cation. Here improvements concerning the thermal stability were made by using the phosphonium based trishexyltetradecylphosphonium $^{6,6,6,14}P^+$ for the model Keggin system, so that the onset of thermal decomposition could be shifted upwards by 105 °C, which drastically improves its applicability for industrial processes. Using other sterically demanding counter ions like imidazol-based cations could possibly expand the thermal window even further, which needs to be researched in further experiments.

However no matter how stable or active a system is, it will only turn into a large scale application if the two basic factors reproducibility and manageable synthesis are ensured. From the results of this work it can already be seen that both aspects are fulfilled for POM-ILs:

1) Synthesis is quite straightforward due to self assembly with short reaction times and easy separation into the organic phase, achieved by pairing the polyoxometalate with the counter ion of choice, such as tetraheptylammonium. This ensures atom efficient synthesis and relative inexpensive production given the complexity of the systems.

2) Cluster syntheses may at many points be hard to predict with pH, temperature, concentration and reaction time having particular effect on the outcome of a reaction. However synthetic routes and reaction principles, once established, are safe roads to travel if it comes to reproducibility. Substitution of the Keggin clusters with the various d-elements used in this work, which were of different size and charge, followed the same synthetic path each time, leading to well defined products.

For these reasons as well as the remarkable ability of polyoxometalates to form extended molecular structures in self assembly processes, the chemistry of the POM-ILs investigated seems to be a highly promising field for future investigation beyond basic research. In this respect the work at hand could help to develop reliable reaction paths for the synthesis of POM-ILs and it established a well-founded basis of the chemo physical properties of Keggin and Wells-Dawson based ionic liquids. These were theoretically underpinned, which may in the future contribute to better understand and ultimately predict POM-IL properties based on their molecular composition. The usefulness of POM-ILs was not only highlighted based on state of the art research in the introduction, it was also successfully applied for CO and high-valent iron chemistry, which can be expected to provide fundamental contributions of POM-IL chemistry to both industrial use and expansion of the boundaries of science.

5.Experimental Section
5.1 Materials

All reagents and chemicals were supplied by SIGMA ALDRICH CHEMICAL COMPANY LTD., FISHER CHEMICALS, ABCR CHEMICALS and ACROS ORGANICS. The materials were used without further purification. Unless described otherwise the reactions were conducted under ambient atmosphere.

5.2 Instrumentation

The following analytical instruments were used for analysis and characterization of the products:

FT-IR spectroscopy: Shimadzu FT-IR-8400S spectrometer. The ionic liquid samples were applied as thin film on the surface of a KBr crystal IR-cell and measured in transmission mode. Intensities are denoted as vs = very strong, s = strong, m = medium, w = weak

EA: Euro Vector Euro EA 3000 elemental analyzer.

UV-Vis-spectroscopy: Shimadzu UV-2401PC spectrophotometer in transmission mode. Quartz cuvettes with 1.0 cm optical path length were used.

^{1}H/^{13}C-NMR: Spectra were recorded with Jeol EX- 270 DELTA und Jeol Lambda- 400 spectrometers. Signals of educts, possible sideproducts and solvents are not mentioned. Chemical shifts are listed according to the δ–scale in ppm and refer to the not deuterated part of the solvents used [$δ_h$ (CDCl$_3$) = 7.27ppm]. Coupling constants are listed in Hertz Hz. For the evaluation of signal multiplicities the following abbreviations are used: s = singlet, d = doublet, t = triplet, m = multiple, dd = doublet of doublets

EPR spectroscopy: JEOL JES-FA200 ESR spectrometer with an X-band microwave unit.

Mössbauer: ^{57}Fe Mössbauer spectra were recorded on a with ElMössbauer spectrometer (MRG-500) at 77 K in constant acceleration mode. ^{57}Co/Rh was used as the radiation source. WinNormos for Igor Pro software has been used for the quantitative evaluation of the spectral parameters (least-squares fitting to Lorentzian peaks). The minimum experimental line widths were 0.20 mms^{-1}. The temperature of the samples was controlled by an MBBC-HE0106 MÖSSBAUER He/N$_2$ cryostat within an accuracy of ±0.3 K. Isomer shifts were determined relative to α-iron at 298 K.

TGA: SETARAM SETSYS Evolution TGA in upstream mode with heating rate of 5 K/min in the temperature area of 30-600 °C, 50 Nml/min, He carrier gas stream and quartz glass vessel (0.5 ml).

MS: UHR-TOF Bruker Daltonik (Bremen, Germany) maXis, an ESI-ToF MS capable of resolution of at least 40,000 FWHM. Detection was in negative ion-mode and the source voltage was 4.5 kV. The flow rates were 180 µl/hour. The drying gas (N$_2$), to aid solvent removal, was held at 180 °C, respectively −20 °C for cryo measurements.

DSC: NETZSCH DSC 204F1, known amount of sample placed in Al pan with pierced lid, gaseous N$_2$ cooling at 100% with flow rate of 20.0 ml/min and 15 K/min. Calibration performed using an indium standard. Temperature range: -100 °C - 60 °C.

5.3 Synthesis and Characterization

As the synthesis of one specific cluster-type e.g. the {SiW$_{11}$M} mainly followed the same synthetic route with (I) the synthesis of the lacunary cluster and (II) the incorporation of a 3d-heteroelement, are the synthetic instructions combined in the following section where appropriate.

5.3.1 Monolacunary Keggin-type Clusters
5.3.1.1 K$_8$[α-SiW$_{11}$O$_{39}$]·13(H$_2$O)

Sodium metasilicate 11 g (50 mmol) was dissolved at room temperature in 100 ml of distilled water (solution A). In a 1-l beaker 182 g (0.55 mol) of sodium tungstate was dissolved in 300 ml of boiling distilled water (solution B). Under vigorous stirring a solution of 4M HCl (165 ml) was added drop wise to the boiling solution B, in ca. 30 min. The local precipitate of tungstic acid, which at first formed with the HCl, was dissolved completely. Solution A was then added, followed by quick addition of further 50 ml of 4M HCl. The pH reading showed a pH of 4.95. The solution was kept boiling for 1 h. After cooling to room temperature, potassium chloride (150 g) was added to the solution and stirred shortly. The white solid product was collected by filtration, washed with two 50-ml portions of a KCl solution (1M) followed by two 30 ml portions of cold water. The white solid obtained was recrystallized from water and the white crystalline product dried in the exsiccator under vacuum for 5 days.

Yield: 109.4 g (34.0 mmol, 67.9% based on Si)

IR (characteristic bands in cm^{-1}): 3420 (s), 2369 (w), 1618 (m), 1120 (w), 997 (m), 961 (s), 893 (s), 797 (s), 727 (s), 538 (m), 511 (m), 474 (m)

5.3.1.2 $K_8[\beta_2\text{-}SiW_{11}O_{39}]\cdot 14H_2O$

Sodium metasilicate 11 g (50 mmol) was dissolved at room temperature in 100 ml of distilled water (solution A). In a 1-l beaker 182 g (0.55 mol) of sodium tungstate was dissolved in 300 ml of distilled water, also at room temperature (Solution B). Under vigorous stirring a solution of 4M HCl (165 ml) was added drop wise to the boiling solution B over 10 min. The local precipitate of tungstic acid, which at first is formed with the HCl was dissolved completely. Solution A was then added, followed by quick addition of further 4M HCl until the pH reading was 5.5. The solution was stirred for 100 min under constant pH adjustment with 4M HCl so that the pH was held constant at 5.5. After cooling to room temperature, potassium chloride (90 g) was added to the solution and stirred for 15 min. The precipitated product was collected by filtration, dried in the exsiccator under vacuum for 3 days and used without further purification to convert it into the {SiW_{10}} cluster (see 5.3.2.1).

Yield: 101.7 g (31.4 mmol, 62.8% based on Si)

5.3.1.3 $K_7[\alpha\text{-}PW_{11}O_{39}]\cdot 9(H_2O)$

40 g (13.9 mmol) of $H_3PW_{12}O_{40}\cdot H_2O$ were dissolved in 200 ml of water. Addition of 2 g of KCl caused a white precipitate to form. The pH was adjusted to pH = 5 with 1M $KHCO_3$ (ca. 60 ml) and the reaction mixture stirred at this pH value and room temperature for 10 min. A light purple precipitate that had formed was centrifuged off and the clear reaction mixture concentrated to half its solvent volume. After 3 d in the refrigerator white crystals of $K_7[\alpha\text{-}PW_{11}O_{39}]\cdot 9(H_2O)$ had formed that where collected by filtration, washed with slight amounts of cool water and dried for 5d under reduced pressure in the exsiccator.

Yield: 30.30 g (9.73 mmol, 70.2 % based on $H_3PW_{12}O_{40}\cdot H_2O$)

5.3.1.4 $(Q^7)_8[\alpha\text{-SiW}_{11}O_{39}]$

A solution of 2.5 g (0.776 mmol) $K_8[\alpha\text{-SiW}_{11}O_{39}]\cdot 13H_2O$ (1.00 eq.) was dissolved in 50 ml of water, heated to 50 °C and a solution of 3.04 g (6.21 mmol) THA-Br (8.00 eq.) in 80 ml of toluene added. The mixture was vigorously stirred for 5 minutes and the organic layer separated. After removal of the solvent toluene the light yellow highly viscous liquid was solvent-stripped once with 50 ml toluene and three times with 50 ml chloroform.

Yield: 1.89 g (3.1710^{-4} mol, 40.9 % based on W)

EA in wt. - % (calculated values in brackets): C 46.28 (45.14), H 8.33 (8.12), N 1.91 (1.88)

IR (characteristic bands in cm^{-1}): 2956/2928/2860 (s), 2463 (w), 1629 (w), 1483/1467 (m), 1379 (m), 1252 (w), 1070/1059 (w), 993 (m), 890 (s), 802 (s), 744 (s), 657 (m), 531 (w)

5.3.1.5 $(Q^5\text{-}Q^8)_n[\alpha\text{-SiW}_{11}O_{39}M(H_2O)]$

Synthesis: A solution of 1.05 eq. of metal salt in form of its chloride or nitrate salt was dissolved in water and added to a solution of 1.00 eq. $K_8[\alpha\text{-SiW}_{11}O_{39}]\cdot 13H_2O$ in water (T = 55 °C). The mixture was stirred for 1.5 hours at 55 °C and the pH slightly acidified with HNO_3 in case of nitrate metal salts or with HCl in case of chloride metal salts. After cooling to room-temperature the tetraalkylammonium counter ion was added in the ratio given in Table 5 as its bromine salt dissolved in toluene and the two-phases mixed vigorously for a short time. Adding the tetraalkylammonium bromide in slightly lower amounts than the stoichiometric ratio resulted in cleaner products according to elemental analysis. The organic layer was then separated and filtered through a folded filter before removing the solvent under reduced pressure. The resulting product was solvent-stripped twice with 50 ml toluene and dried under vacuum at 60 °C for at least 24 h and with several times lyophilizing.

Entry No	compound formula	Eq. $(Q^y)Br$ [a]	Yield [%]	color
1	$(Q^5)_6[\alpha\text{-}SiW_{11}O_{39}Cu(H_2O)]$	5.9	54	spring green
2	$(Q^5)_5[\alpha\text{-}SiW_{11}O_{39}Fe(H_2O)]$	4.9	67	goldenrod
3	$(Q^6)_6[\alpha\text{-}SiW_{11}O_{39}Cu(H_2O)]$	5.9	64	spring green
4	$(Q^6)_5[\alpha\text{-}SiW_{11}O_{39}Fe(H_2O)]$	4.9	65	goldenrod
5	$(Q^7)_6[\alpha\text{-}SiW_{11}O_{39}Cu(H_2O)]$	5.9	75	spring green
6	$(Q^7)_6[\alpha\text{-}SiW_{11}O_{39}Co(H_2O)]$	5.9	64	dark red
7	$(Q^7)_6[\alpha\text{-}SiW_{11}O_{39}Ni(H_2O)]$	5.9	65	yellow
8	$(Q^7)_6[\alpha\text{-}SiW_{11}O_{39}Mn(H_2O)]$	5.9	67	purple
9	$(Q^7)_5[\alpha\text{-}SiW_{11}O_{39}Fe(H_2O)]$	4.9	70	brown
10	$(Q^7)_5[\alpha\text{-}SiW_{11}O_{39}Cr(H_2O)]$	4.9	71	dark green
11	$(Q^8)_6[\alpha\text{-}SiW_{11}O_{39}Cu(H_2O)]$	5.9	65	lime green
12	$(Q^8)_5[\alpha\text{-}SiW_{11}O_{39}Fe(H_2O)]$	4.9	76	goldenrod

[a] A marginally sub-stoichiometric amount of the respective alkylammonium bromide was employed in the syntheses as this leads to improved purity of the final product (based on elemental analysis).

Table 11 Summary of synthesized compounds with the respective yield and equivalents of tetraalkylammonium bromide added.

EA in wt. - % (calculated values in brackets):

Entry No	compound formula	C	H	N
1	$(Q^5)_6[\alpha\text{-}SiW_{11}O_{39}Cu(H_2O)]$	31.36 (31.82)	5.96 (5.88)	1.72 (1.86)
2	$(Q^5)_5[\alpha\text{-}SiW_{11}O_{39}Fe(H_2O)]$	31.36 (28.44)	5.85 (5.25)	1.65 (1.66)
3	$(Q^6)_6[\alpha\text{-}SiW_{11}O_{39}Cu(H_2O)]$	35.66 (35.41)	6.49 (6.48)	1.65 (1.72)
4	$(Q^6)_6[\alpha\text{-}SiW_{11}O_{39}Fe(H_2O)]$	31.58 (32.00)	5.79 (5.82)	1.42 (1.56)
5	$(Q^7)_6[\alpha\text{-}SiW_{11}O_{39}Cu(H_2O)]$	39.61 (38.78)	7.16 (6.97)	1.69 (1.62)
6	$(Q^7)_6[\alpha\text{-}SiW_{11}O_{39}Co(H_2O)]$	38.86 (38.82)	6.98 (6.98)	1.68 (1.62)

7	$(Q^7)_6[\alpha\text{-SiW}_{11}O_{39}Ni(H_2O)]$	39.35 (38.82)	7.08 (6.98)	1.49 (1.62)
8	$(Q^7)_6[\alpha\text{-SiW}_{11}O_{39}Mn(H_2O)]$	36.41 (38.85)	6.48 (6.99)	1.46 (1.62)
9	$(Q^7)_5[\alpha\text{-SiW}_{11}O_{39}Fe(H_2O)]$	36.98 (35.14)	6.61 (6.32)	1.45 (1.46)
10	$(Q^7)_5[\alpha\text{-SiW}_{11}O_{39}Cr(H_2O)]$	37.08 (35.18)	6.65 (6.33)	1.72 (1.47)
11	$(Q^8)_6[\alpha\text{-SiW}_{11}O_{39}Cu(H_2O)]$	41.51 (41.50)	7.42 (7.44)	1.51 (1.51)
12	$(Q^8)_5[\alpha\text{-SiW}_{11}O_{39}Fe(H_2O)]$	39.39 (37.81)	6.96 (6.78)	1.43 (1.38)

Table 12 Summary of the elemental analysis of all tetraalkylammonium undecatungstosilicates given for C,H and N in weight percent. Calculated values are given in brackets.

^{1}H- and ^{13}C-NMR of the tetraheptylammonium cations:

^{1}H-NMR (chloroform-D, 400 MHz): δ [ppm] = 2.33 (2H; H-1), 1.32 (10H; H-2 - H-6), 0.88 (3H; H-7)

^{13}C-NMR (chloroform-D, 400 MHz): δ δ [ppm] = 58.92 (C-1), 31.59 (C-5), 28.90 (C-4), 26.26 (C-2), 22.63 (C-6), 22.30 (C-3), 14.13 (C-7)

IR: For the most important bands see 3.1.3.1

UV-Vis spectra:

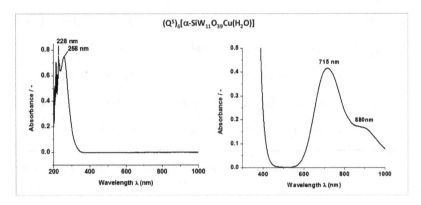

Figure 49 UV-Vis spectrum of $(Q^5)_6[\alpha\text{-SiW}_{11}O_{39}Cu(H_2O)]$ in CH_2Cl_2. Left: c = 18.0 µM, the extinction coefficients are ε_{228} = 41445 $M^{-1}cm^{-1}$ and ε_{258} = 41581 $M^{-1}cm^{-1}$. Right: c = 18.0 mM, the extinction coefficients are ε_{715} = 23.1 $M^{-1}cm^{-1}$ and ε_{880} = 9.3 $M^{-1}cm^{-1}$.

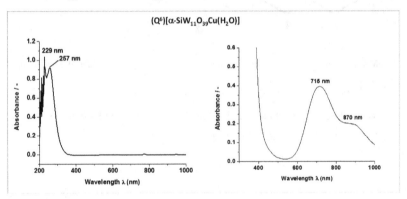

Figure 50 $(Q^6)_6[\alpha\text{-SiW}_{11}O_{39}Cu(H_2O)]$ in CH_2Cl_2. Left: c = 35.3 µM, the extinction coefficients are ε_{229} = 27748 $M^{-1}cm^{-1}$ and ε_{257} = 26080 $M^{-1}cm^{-1}$. Right: c = 17.6 mM, the extinction coefficients are ε_{715} = 22.4 $M^{-1}cm^{-1}$ and ε_{870} = 11.3 $M^{-1}cm^{-1}$.

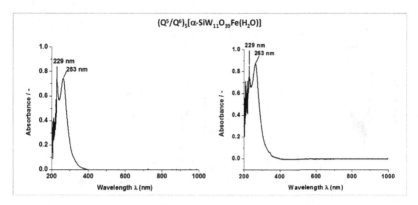

Figure 51 Left: UV-Vis spectrum of $(Q^5)_5[\alpha\text{-SiW}_{11}O_{39}Fe(H_2O)]$ in CH_2Cl_2 (c = 28.5 μM). The extinction coefficients are $\varepsilon_{229} = 26808\ M^{-1}cm^{-1}$ and $\varepsilon_{263} = 26100\ M^{-1}cm^{-1}$. Right: UV-Vis spectrum of $(Q^6)_5[\alpha\text{-SiW}_{11}O_{39}Fe(H_2O)]$ in CH_2Cl_2 (c = 18.2 μM). The extinction coefficients are $\varepsilon_{229} = 44674\ M^{-1}cm^{-1}$ and $\varepsilon_{263} = 47786\ M^{-1}cm^{-1}$.

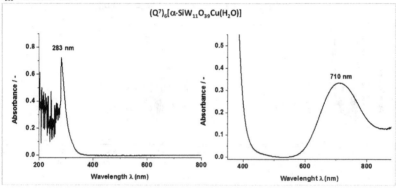

Figure 52 UV-Vis spectra of $(Q^7)_6[\alpha\text{-SiW}_{11}O_{39}Cu(H_2O)]$ in toluene. left: c = 4.5 10^{-5}M, the extinction coefficient is $\varepsilon_{283} = 16165\ M^{-1}cm^{-1}$. Right: c = 17.9 mM, the extinction coefficient is $\varepsilon_{710} = 18.6\ M^{-1}cm^{-1}$.

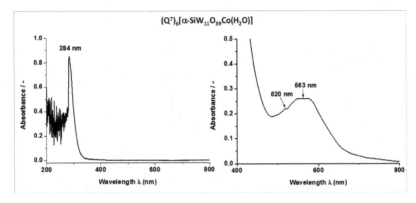

Figure 53 UV-Vis spectrum of $(Q^7)_6[\alpha\text{-SiW}_{11}O_{39}Co(H_2O)]$ in toluene. Left: c = 2.3 10^{-5}M, the extinction coefficient is ε_{284}= 36731 $M^{-1}cm^{-1}$. Right: c = 3.85 mM, the extinction coefficients are ε_{520} = 56.9 $M^{-1}cm^{-1}$ and ε_{563} = 67.8 $M^{-1}cm^{-1}$.

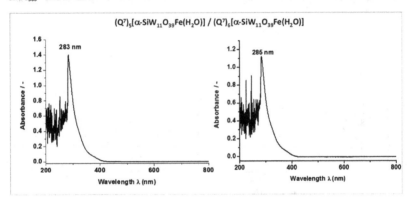

Figure 54 Left: UV-Vis spectrum of $(Q^7)_5[\alpha\text{-SiW}_{11}O_{39}Fe(H_2O)]$ in toluene (c = 6.5 10^{-5}M). The extinction coefficient is ε_{283} = 21636 $M^{-1}cm^{-1}$. UV-Vis spectrum of $(Q^7)_6[\alpha\text{-SiW}_{11}O_{39}Fe(H_2O)]$ in toluene (c = 6.2 10^{-6} M). The extinction coefficient is ε_{285} = 18198 $M^{-1}cm^{-1}$.

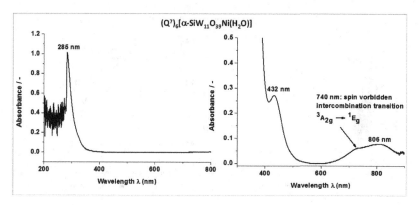

Figure 55 UV-Vis spectrum of $(Q^7)_6[\alpha\text{-SiW}_{11}O_{39}Ni(H_2O)]$ in toluene. Left: c = 6.4 10^{-5}M), the extinction coefficient is ε_{285} = 15704 $M^{-1}cm^{-1}$. Right: c = 10.4 mM, the extinction coefficients are ε_{432} = 26.1 $M^{-1}cm^{-1}$ and ε_{806} = 7.5 $M^{-1}cm^{-1}$.

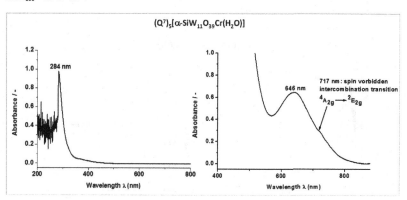

Figure 56 UV-Vis spectrum of $(Q^7)_5[\alpha\text{-SiW}_{11}O_{39}Cr(H_2O)]$ in toluene. Left: c = 3.7 10^{-5}M, the extinction coefficient is ε_{284} = 26012 $M^{-1}cm^{-1}$. Right: c = 20.4 mM, the extinction coefficient is ε_{646} = 31.3 $M^{-1}cm^{-1}$.

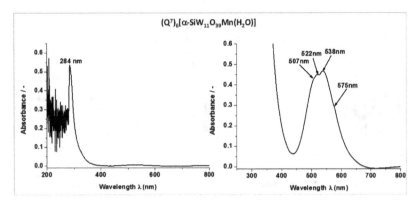

Figure 57 UV-Vis spectrum of $(Q^7)_6[\alpha\text{-}SiW_{11}O_{39}Mn(H_2O)]$ in toluene. Left: c = 2.7 10^{-5} M, the extinction coefficient is ε_{284} = 19819 $M^{-1}cm^{-1}$. Right: c = 1.62 mM, the extinction coefficients are ε_{507} = 265.0 $M^{-1}cm^{-1}$, ε_{522} = 278.4 $M^{-1}cm^{-1}$, ε_{538} = 289.4 $M^{-1}cm^{-1}$ and ε_{575} = 180.3 $M^{-1}cm^{-1}$.

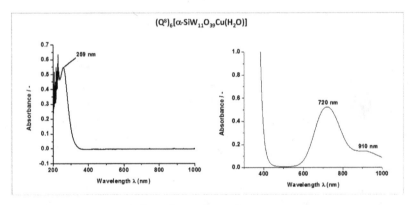

Figure 58 UV-Vis spectrum of $(Q^8)_6[\alpha\text{-}SiW_{11}O_{39}Cu(H_2O)]$ in CH_2Cl_2. Left: c =18.2 µM, the extinction coefficient is ε_{259} = 30145 $M^{-1}cm$. Right: c =18.2 mM, the extinction coefficients are ε_{720} = 28.8 $M^{-1}cm^{-1}$ and ε_{910} = 7.8 $M^{-1}cm^{-1}$.

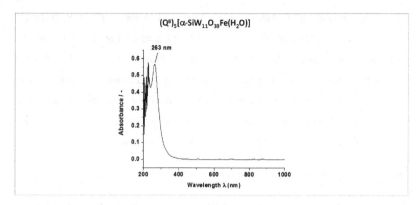

Figure 59 UV-Vis spectrum of $(Q^8)_5[\alpha\text{-SiW}_{11}O_{39}Fe(H_2O)]$ in CH_2Cl_2 (c = 9.3 µM). The extinction coefficient is $\varepsilon_{263} = 58976 \ M^{-1}cm^{-1}$.

5.3.1.6 $(Q^7)_{4/5}[\alpha\text{-PW}_{11}O_{39}M(H_2O)]$ M = Cu^{II}/Fe^{III}

For the synthesis of $(Q^7)_5[\alpha\text{-PW}_{11}O_{39}Cu(H_2O)]$ an excess of $CuCl_2 \cdot 2H_2O$ (1.00 g, 5.87 mmol) was dissolved in 40 ml of water, together with 5.00 g (15.16 mmol) $Na_2WO_4 \cdot 2H_2O$. After addition of 40 ml of 1M H_3PO_4 a temporary blue precipitate formed which completely dissolved again after stirring the reaction mixture for 10 minutes. Afterwards 4.3 eq. of Q^7-Br (2.94 g, 6.00 mmol) in 60 ml of toluene were added. In the biphasic system the organic layer turned orchid at first and lime green after shaking in the separation funnel. The organic layer was separated solvent stripped once with 50 ml toluene and the solvent removed partially. The product then formed a separate dark lime green colored phase as the bottom layer with the aqueous layer as light lime green colored top layer. Under vacuum the biphasic system was completely removed of solvent and dried there for 3 d with several times lyophilizing.

Yield: 2.627 g (0.55 mmol, 39.6% based on $Na_2WO_4 \cdot 2H_2O$)

EA in wt. - % (calculated values in brackets): C 33.09 (34.94), H 5.98 (6.32), N 1.23 (1.46)

IR (characteristic bands in cm^{-1}): 3449 (m), 2955/2928/2857 (s), 1481 (m), 1460 (m), 1096 (m), 1070 (w), 1054 (w), 1026 (m), 962 (s), 878 (m), 820 (m), 798 (w), 750 (w), 662 (w), 606 (w), 548 (w), 529 (w)

For the synthesis of $(Q^7)_4[\alpha\text{-PW}_{11}O_{39}Fe(H_2O)]$ 5.00 g (1.61 mmol) of $K_7[PW_{11}O_{39}]\cdot9H_2O$ were dissolved in 100 ml of warm water (T = 60 °C). 0.456g (1.69 mmol, 1.05 eq.) of $FeCl_3\cdot6H_2O$ were added in 20 ml of water and the reaction mixture stirred at 60 °C for 1 h. For the phase transfer 3.10 g (6.34 mmol, 3.95 eq.) of Q^7Br were dissolved in 80 ml of toluene and added to the cooled down reaction mixture. The organic layer was separated, the solvent removed under reduced pressure and the yellow highly viscous product dried under vacuum for 24 h with several times lyophilizing.

Yield: 2.627 g (0.55 mmol, 39.6% based on $Na_2WO_4\cdot2H_2O$)

EA in wt. - % (calculated values in brackets): C 31.56 (30.61), H 5.68 (5.55), N 1.23 (1.28)

IR (characteristic bands in cm^{-1}): 3471 (m), 2955/2928/2857 (s), 1481 (m), 1460 (m), 1379 (m), 1069 (m), 962 (s), 885 (m), 812 (m), 725 (w), 671 (w), 662 (w), 594 (w), 517 (w)

5.3.1.7 $(Q^7)_{6/7}[\alpha\text{-BW}_{11}O_{39}M(H_2O)]$ M = Cu^{II}/Fe^{III}

For the synthesis of $(Q^7)_7[\alpha\text{-BW}_{11}O_{39}Cu(H_2O)]$ the $[\alpha\text{-BW}_{11}O_{39}Cu(H_2O)]$ cluster was isolated as $K_7[BW_{11}O_{39}Cu(H_2O)]$ prior to its phase transfer in a second step. For the formation of the cluster 18.15 g (0.055 mol) $Na_2WO_4\cdot2H_2O$ were dissolved in 90 ml H_2O and the solutions pH adjusted to 6.3 with HNO_3. 1.50 g (0.024 mol) of H_3BO_3 were added as solid and the solution heated to 85 °C. After addition of 1.21 g (0.005 mol) $Cu(NO_3)_2\cdot3H_2O$ the reaction mixture was stirred for 10 minutes at 85 °C which formed a green precipitate that slowly dissolved upon further heating at 85°C for 10 min. The warm mixture was filtered through a folded paper filter and allowed to stand at room temperature for two hours which caused the formation of light green crystals. These were filtered off and the mother liquor salted out with 10 g of KCl. The resulting turquoise-green precipitate was filtered off, washed with water and dried in the exsiccator under reduced pressure for 5 days.

Yield $K_7[BW_{11}O_{39}Cu(H_2O)]$: 9.46 g (3.14 mmol, 29% based on $Na_2WO_4\cdot2H_2O$)

For the phase transfer 5.00 g (1.66 mmol) of the $K_7[BW_{11}O_{39}Cu(H_2O)]$ were dissolved in 120 ml of warm water (T = 55 °C) and a solution of 5.54 g(11.29 mmol; 6.8 eq.) Q^7-Br in 100 ml toluene added. The organic layer was separated, the solvent removed under reduced pressure and the green highly viscous product dried under vacuum for 24 h with several times lyophilizing.

Yield (Q^7)$_7$[BW$_{11}$O$_{39}$Cu(H$_2$O)]: 7.31 g (1.30 mmol, 78% based on $K_7[BW_{11}O_{39}Cu(H_2O)]$)

EA in wt. - % (calculated values in brackets): C 43.69 (41.93), H 7.95 (7.58), N 1.79 (1.75)

UV-Vis spectrum:

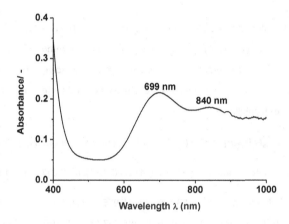

Figure 60 UV-Vis spectrum of $(Q^7)_7[BW_{11}O_{39}Cu(H_2O)]$ in toluene (c = 19.3 mM). The extinction coefficients are ε_{699} = 12 $M^{-1}cm$ and ε_{840} = 10 $M^{-1}cm$.

IR (characteristic bands in cm^{-1}): 3568 (m), 2992/2955/2924/2870/2851 (s), 1638 (m), 1466 (m), 1456 (m), 1377 (m), 1082 (w), 1032 (w), 986 (m), 937 (m), 908 (m), 881 (m), 816 (m), 791 (m), 738 (w), 721 (m),

The $(Q^7)_6[BW_{11}O_{39}Fe(H_2O)]$ was synthesized in a similar fashion. At first $Na_6[BW_{11}O_{39}Fe(H_2O)]$ was synthesized as followed: 18.15 g (0.055 mol) $Na_2WO_4·2H_2O$ were dissolved in 200 ml H_2O and the solutions pH adjusted to 7.0 with HCl (4M). After heating to 80 °C 2.75 g (0.044 mol) of H_3BO_3 were added as solid. The reaction was stirred for 15 min at this temperature and then added to a solution of 2.04 g (0.005 mol)

Fe(NO$_3$)$_3$·9H$_2$O in 25 ml of hot water (T = 85 °C). The pH was again adjusted with HCl to 7.00 and the reaction mixture stirred for 10 minutes at 85 °C which formed a sand-colored precipitate that slowly dissolved. The warm mixture was filtered through a folded paper filter and allowed to cool to room temperature. Addition of 300 ml of ethanol precipitated a brown solid that was separated by centrifugation washed with slight amounts of ethanol and water. It was dissolved in 80 ml of lukewarm water, precipitated a second time with addition of 150 ml of ethanol and the product dried in the exsiccator under reduced pressure for 5 days. The product was obtained as a red-brown solid.

Yield Na$_6$[BW$_{11}$O$_{39}$Fe(H$_2$O)]: 7.45 g (2.44 mmol, 48.8% based on Na$_2$WO$_4$·2H$_2$O)

For the phase transfer 3.55 g (1.16 mmol) of the Na$_6$[BW$_{11}$O$_{39}$Fe(H$_2$O)] were dissolved in 80 ml of warm water (T = 40 °C) and a solution of 3.36 g(6.84 mmol; 5.9 eq.) Q^7-Br in 120 ml toluene added. The organic layer was separated, the solvent removed under reduced pressure and the yellow highly viscous product dried under vacuum for 24 h with several times lyophilizing.

Yield (Q^7)$_6$[BW$_{11}$O$_{39}$Fe(H$_2$O)]: 2.45 g (0.47 mmol, 41% based on Na$_6$[BW$_{11}$O$_{39}$Fe(H$_2$O)])

EA in wt. - % (calculated values in brackets): C 43.04 (38.84), H 7.82 (7.02), N 1.64 (1.62)

IR (characteristic bands in cm^{-1}): 3483 (m), 2992/2955/2924/2870/2851 (s), 1638 (m), 1466 (m), 1456 (m), 1377 (m), 1082 (w), 1032 (w), 947 (m), 879 (m), 779 (m), 721 (m), 671 (w), 540 (w)

5.3.1.8 (^{6,6,6,14}P)$_{5/6}$[α-SiW$_{11}$O$_{39}$M(H$_2$O)]; M = Fe/Cu

Both the (^{6,6,6,14}P)$_6$[α-SiW$_{11}$O$_{39}$Cu(H$_2$O)] (**20**) and (^{6,6,6,14}P)$_5$[α-SiW$_{11}$O$_{39}$Fe(H$_2$O)] (**21**) were synthesized starting with 1.00 eq. of {α-SiW$_{11}$} being dissolved in water (ca. 20 ml/g of cluster) and addition of 1.05 eq. of the hetero d-metal in form of its chloride salt dissolved in water so CuCl$_2$·2H$_2$O is used for (**20**) and FeCl3·6H$_2$O for (**21**). The reaction mixture was stirred at 60°C for 1.5 h and the pH slightly acidified with HCl to 4.5 in case of (20). With the iron compound no additional acidification was necessary. After cooling to room-temperature the trishexyltetradecylphosphonium-counter ion $(^{6,6,6,14}P)^+$ was added as $(^{6,6,6,14}P)Br$ in slightly lower amounts than the stoichiometric ratio. The organic layer was then separated and filtered through a folded filter before removing the solvent under reduced pressure. The resulting yellow highly viscous product was solvent-stripped twice with 50 ml toluene and dried under vacuum at 60 °C for around 24 h and with several times lyophilizing.

Yield: $(^{6,6,6,14}P)_5[\alpha\text{-SiW}_{11}O_{39}Fe(H_2O)]$: 85.9% based on $\{\alpha\text{-SiW}_{11}\}$); $(^{6,6,6,14}P)_6[\alpha\text{-SiW}_{11}O_{39}Cu(H_2O)]$: 76% based on $\{\alpha\text{-SiW}_{11}\}$)

EA in wt. - % (calculated values in brackets):

$(^{6,6,6,14}P)_5[\alpha\text{-SiW}_{11}O_{39}Fe(H_2O)]$: C 36.94 (37.19), H 6.62 (6.67)

$(^{6,6,6,14}P)_6[\alpha\text{-SiW}_{11}O_{39}Cu(H_2O)]$: C 43.54 (40.75), H 7.96 (7.30)

IR (characteristic bands in cm^{-1}):

$(^{6,6,6,14}P)_5[\alpha\text{-SiW}_{11}O_{39}Fe(H_2O)]$: 3458 (m), 2955/2924/2855 (s), 1460 (m), 1408 (w), 1379 (w), 1300 (w), 1263 (w), 1215 (w), 1177 (w), 1111 (w), 1094 (w), 999 (m), 953 (m), 907 (s), 800 (s), 743 (w), 723 (w), 692 (w), 540 (w), 526 (w)

$(^{6,6,6,14}P)_6[\alpha\text{-SiW}_{11}O_{39}Cu(H_2O)]$: 3458 (m), 2955/2928/2857 (s), 1460 (m), 1408 (m), 1379 (w), 1300 (w), 1263 (w), 1215 (w), 1177 (w), 1111 (w), 1093 (w), 999 (m), 953 (m), 907 (s), 800 (s), 743 (w), 700 (m), 604 (w), 723 (w), 692 (w), 540 (w), 527 (w)

5.3.1.9 (OMIM)$_n$[α-SiW$_{11}$O$_{39}$M(H$_2$O)]; M = Cr

A solution of 0.39 g (0.98 mmol, 1.05 eq.) of Cr(NO$_3$)3 ·H2O was dissolved in water and added to a solution of 3.00 g (0.93 mmol, 1.00 eq.) K$_8$[α-SiW$_{11}$O$_{39}$]·13H$_2$O in water (T = 55 °C). The mixture was stirred for 1.5 hours at 55 °C and the pH slightly acidified to pH = 6 with HNO$_3$ (1M). After cooling to room-temperature 1.05 g (4.56 mmol, 4.9 eq.) of 1-methyl-3-octylimidazolium chloride OMIM-Cl dissolved in 100 ml toluene were added and the two-phases mixed vigorously for a short time which resulted in the precipitation of a green solid. It was isolated by sedimentation and decantation of the reaction mixture followed by thorough washing with toluene and water respectively. It was dried under reduced pressure in the exsiccator for 5d.

Yield: 2.10 g (0.56 mmol, 60% based on $K_6[BW_{11}O_{39}Fe(H_2O)]$)

EA in wt. - % (calculated values in brackets): C 18.52 (19.37), H 2.99 (3.17), N 3.56 (3.76)

5.3.2 Dilacunary Keggin-type Clusters
5.3.2.1 $K_8[\gamma\text{-SiW}_{10}O_{36}]\cdot12H_2O$

45 g (15 mmol) of the β_2-isomer of undecatungstosilicate $K_8[\beta_2\text{-SiW}_{11}O_{39}]$ $14H_2O$ synthesized as described under 5.3.1.2 was dissolved in 450 ml of water at room temperature. Undissolved impurities of paratungstate were rapidly removed via filtration through Celite. The pH of the solution was quickly adjusted to 9.1 by addition of a 2M aqueous solution of K_2CO_3 and kept at this pH by further addition of the K_2CO_3 solution for exactly 16 min under stirring. Afterwards 120 g of KCl were added and the solution stirred for 10 min under continued pH control. Precipitated product of the γ-decatungstosilicate was then collected by filtration and washed with two portions of 30 ml aqueous 1M KCl solution. The light yellow crystalline product was dried under reduced pressure in the exsiccator.

Yield: 17.2 g (5.79 mmol, 38.6% based on $K_8[\beta_2\text{-SiW}_{11}O_{39}]$)

IR (characteristic bands in cm^{-1}): 3446 (s), 2362/2341 (w), 1622 (w), 988 (w), 943 (m), 905 (m), 866 (s), 818 (m), 745 (s), 667 (w), 650 (w), 530 (w)

5.3.2.2 $(Q^7)_4[\alpha\text{-SiW}_{10}O_{36}Cu_2(H_2O)_2]$

A solution of 3.00 g (0.97 mmol, 1.00 eq.) of $K_8[\gamma\text{-SiW}_{10}O_{36}]$ was dissolved in 50 ml of water and 0.347 g (2.03 mmol, 2.1 eq.) of $CuCl_2\cdot2H_2O$ added in 20 ml of water. The pH reading was 3.50. The reaction mixture was stirred for 1h at a temperature of 55 °C. After cooling to room temperature 1.85 g (3.78 mmol, 3.90 eq.) Q^7-Br dissolved in

100 ml toluene were added and the two-phases mixed vigorously for a short time. The organic layer was then separated and filtered through a folded filter before removing the solvent under reduced pressure. The resulting yellow highly viscous product was dried under vacuum at 60 °C for around 24 h and with several times lyophilizing.

Yield: 2.49 g (0.59 mmol, 60% based on $K_8[\gamma\text{-SiW}_{10}O_{36}]$)

EA in wt. - % (calculated values in brackets): C 35.02 (31.66), H 6.54 (5.79), N 1.41 (1.32)

IR (characteristic bands in cm^{-1}): 3485 (m), 2955/2928/2857 (s), 1481 (m), 1458 (m), 1379 (m), 1001 (m), 961 (m), 907 (s), 806 (s), 725 (m), 708 (m), 662 (w), 548 (w), 529 (w)

UV-Vis spectra:

$(Q^7)_4[\alpha\text{-SiW}_{10}O_{36}Cu_2]$

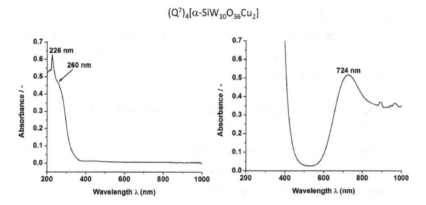

Figure 61 UV-Vis spectrum of $(Q^7)_4[\alpha\text{-SiW}_{10}O_{36}Cu_2(H_2O)_2]$ in CH_2Cl_2. Left: c = 19.0 µM, the extinction coefficient is ε_{226} = 32674 M^{-1}cm^{-1}. Right: c = 19.0 mM, the extinction coefficient is ε_{724} = 27 M^{-1}cm^{-1}.

5.3.3 The Dawson-type Clusters
5.3.3.1 $\alpha/\beta\text{-}K_6[P_2W_{18}O_{62}]\cdot 19H_2O$ (+ $K_{14}NaP_5W_{30}O_{110}$)

In the course of this synthesis the use of metal spatulas was avoided as these may reduce the clusters!

250 g (0.76 mol) of $Na_2WO_4\cdot 2H_2O$ were dissolved in 450 ml of water and 210 ml (3.09 mol) of orthophosphoric acid (85%) added slowly. The solution was heated at

reflux for 4 h at an oil-bath temperature of 130 °C. After cooling to room temperature, 100 g of ammonium chloride were added, and the solution stirred for 10 min. The pale yellow precipitate was separated by filtration, dissolved again in 600 ml of water and precipitates with a further portion of l00 g ammonium chloride. After stirring for 10 min and filtration through a course frit, and suction, the precipitate was dissolved in 250 ml of warm water (ca. 45°C and left to evaporate at room temperature for 6 days until the crystallization of the b-isomer ammonium salt was almost complete. The green crystals were collected by filtration and vacuum dried for 2 days.

Yield of $K_6[\beta_2-P_2W_{18}O_{62}]\cdot 19H_2O$: 22.2 g (4.49 mmol, 10.6% based on $Na_2WO_4\cdot 2H_2O$)

To the filtrate of the crystallization a 40 g (0.54 mol) quantity of potassium chloride was added, stirred for 15 min and the precipitate collected on a filter. Dissolving it in 250 ml of hot water (ca. 80 °C) and slow cooling to room temperature yielded 4.93 g (0.616 mmol) of white needles ($K_{14}Na\ P_5W_{30}O_{110}$) which were collected by filtration after a crystallization time of 5 hours.

Yield of $K_{14}Na\ P_5W_{30}O_{110}$: 4.93 g (0.616 mmol, 2.4% based on $Na_2WO_4\cdot 2H_2O$)

The filtrate was heated to ebullition and after cooling to room temperature a 25-g quantity of potassium chloride added. The crude a-octadecatungstodiphosphat was then collected on a filter and air dried under vacuum for 2 days.

Yield of $K_6[\alpha_2-P_2W_{18}O_{62}]\cdot 19H_2O$: 127.3 g (25.8 mmol, 61.0% based on $Na_2WO_4\cdot 2H_2O$)

5.3.3.2 $(Q^7)_6[\alpha_2-P_2W_{18}O_{62}]$

The $K_6[\beta_2-P_2W_{18}O_{62}]\cdot 19H_2O$ was directly phase transferred after its isolation. For this 1.50 g (0.326 mmol) $K_6[\alpha_2-P_2W_{18}O_{62}]\cdot 19H_2O$ (1.00 eq.) were dissolved in 20 ml of water and a solution of 0.960 g (1.96 mmol) THA-Br (6.00 eq.) in 30 ml of toluene added. The mixture was stirred for 10 minutes and the organic layer separated. After removing the solvent the product was obtained as white solid, that was solvent-stripped once with 30 ml toluene and three times with 30 ml chloroform.

Yield: 2.09 g (0.34 mmol, 93.9% based on{α–W_{18}})

EA in wt. - % (calculated values in brackets): C 30.14 (29.55), H 5.41 (5.31), N 1.16 (1.23)

5.3.3.3 $K_{10}[\alpha_2\text{-}P_2W_{17}O_{61}]\cdot20H_2O$

A portion of 80.00 g (0.015 mol) $K_6[\alpha_2\text{-}P_2W_{18}O_{62}]\cdot19H_2O$ (1.00 eq.) was dissolved in 200 ml of water and a solution of 20.00 g (0.20 mol) $KHCO_3$ in 200 ml of water added. The mixture was stirred at room temperature for 2 hours and the white precipitate filtered of. It was recrystallized in 500 ml of hot water (T = 95 °C) which yielded white crystals after 4 hours. These were filtered of and dried under vacuum.

Yield: 47.59 g (9.68 mmol, 84.2% based on{α–W_{18}})

5.3.3.4 $(Q^7)_8[\alpha_2\text{-}P_2W_{17}O_{61}Co(H_2O)]$

5.00 g (1.02mmol) $K_{10}[a_2\text{-}P_2W_{17}O_{61}]\cdot20H_2O$ (1.00 eq.) were dissolved in 50 ml of water and a solution of 0.30 g (1.02mmol) $Co(NO_3)_2\cdot6H_2O$(1.00 eq.) in 15 ml of water were added slowly. The pH was adjusted to 3.5 with HNO_3 and the reaction mixture was heated to 95 °C and stirred for 1 hour. After cooling to room temperature 4.00 g (8.16 mmol) THA-Br (8 eq.) in 50 ml toluene were added and stirred vigorously for a short time. The dark red organic layer was separated and filtered through a folded filter. After removing the solvent the product was solvent-stripped once with 50 ml toluene and three times with 50 ml chloroform. The brown, highly viscous product was dried for one day under vacuum at 55°C.

Yield: 4.33 g (0.86 mmol, 84.6% based on {α–W_{17}})

EA in wt. - % (calculated values in brackets): C 36.19 (35.80), H 6.53 (6.44), N 1.43 (1.49)

IR (characteristic bands in cm^{-1}): 3507 (m), 2955/2924/2870/2851 (s), 1467 (m), 1456 (m), 1383 (m), 1342 (m), 1086 (s), 1013 (w), 945 (m), 916 (m), 814 (s), 768 (w), 723 (m), 600 (w), 565 (w), 529 (w)

5.3.4 Azide-experiment $(Q^7)_5[\alpha-SiW_{11}O_{39}Fe(H_2O)]$ + NaN$_3$

0.677 g (0.14 mmol) of $(Q^7)_5[SiW_{11}O_{39}Fe(H_2O)]$ were dissolved in 25 ml of dry chloroform and 0.055 g (0.84 mmol; 6 eq.) of NaN$_3$ added under nitrogen atmosphere applying normal Schlenk technique. The reaction mixture was stirred at 45 °C for 4 h. Not reacted sodium azide was filtered off and the bright yellow solution split into two parts (ca. 50% each). One part was directly freed from solvent and dried at the vacuum to yield a yellow highly viscous liquid.

The other part was transferred into an EPR quartz tube and photolyzed for 24 h with an 150 W medium-pressure mercury light source (Heraeus TQ-150, UV Consulting Peschl λ_{max} = 365 nm equipped with a pyrex deep-UV cutoff filter λ_{cutoff} = 320 nm) which resulted in a dark blue solution that was also freed from the solvent and dried at the vacuum.

5.3.5 CO-experiments

A concentrated solution of $(Q^7)_{5/6}(\alpha-SiW_{11}O_{39}M(H_2O))$ (M = Cu^{II}, Mn^{II}, Fe^{III}) as well as $(Q^7)_8(\alpha-SiW_{11}O_{39})$ in toluene was prepared in a schlenk tube sealed from air via septum. Utilizing a syringe a constant stream of CO was bubbled for 5 hours through the solution under vigorous stirring. The following color changes occurred: $(Q^7)_6(\alpha-SiW_{11}O_{39}Cu(H_2O))$: turquoise to lime green. $(Q^7)_6(\alpha-SiW_{11}O_{39}Mn(H_2O))$: purple to orange. $(Q^7)_5(\alpha-SiW_{11}O_{39}Fe(H_2O))$: no visual color change. $(Q^7)_8(\alpha-SiW_{11}O_{39})$: white to khaki.

Afterwards the solvent was removed and the highly viscous product dried at the vacuum for 3 days and several times lyophilizing.

6. References:

1. H. Davy, *Biggs and Cottle*, **1800**, 1800.

2. C. Reichardt, *Organic Process Research & Development*, **2007**, *11*, 105-113.

3. W. Sundermeyer, *Angew. Chem. Int. Ed.*, **1965**, *4*, 222-238.

4. T. Torimoto, T. Tsuda, K. Okazaki and S. Kuwabata, *Advanced Materials*, **2010**, *22*, 1196-1221.

5. F. E. M. Armand, D. R. MacFarlane, H. Ohno, B. Scrosati, *Nat.Mater.*, **2009**, *8*, 621-629.

6. S. Werner, M. Haumann and P. Wasserscheid, *Annual Review of Chemical and Biomolecular Engineering*, **2010**, *1*, 203-230.

7. H. Olivier-Bourbigou, L. Magna and D. Morvan, *Appl. Catal. A*, **2010**, *373*, 1-56.

8. T. Welton, *Chem. Rev.*, **1999**, *99*, 2071-2084.

9. J. P. Hallett and T. Welton, *Chem. Rev.*, **2011**, *111*, 3508-3576.

10. K. R. S. Martyn J. Earle, *Pure Appl. Chem.*, **2000**, *72*, 1391-1398.

11. T. Welton, P. Wasserscheid, Ionic Liquids in Synthesis, *Wiley-VCH*, **2007**.

12. M. C. Buzzeo, R. G. Evans and R. G. Compton, *ChemPhysChem*, **2004**, *5*, 1106-1120.

13. M. Galiński, A. Lewandowski and I. Stępniak, *Electrochim. Acta*, **2006**, *51*, 5567-5580.

14. B. Smarsly and H. Kaper, *Angew. Chem. Int. Ed.*, **2005**, *44*, 3809-3811.

15. A. I. Bhatt, A. M. Bond, D. R. MacFarlane, J. Zhang, J. L. Scott, C. R. Strauss, P. I. Iotov and S. V. Kalcheva, *Green Chem.*, **2006**, *8*, 161-171.

16. J. Dupont, R. F. de Souza and P. A. Z. Suarez, *Chem. Rev.*, **2002**, *102*, 3667-3692.

17. V. I. Pârvulescu and C. Hardacre, *Chem. Rev.*, **2007**, *107*, 2615-2665.

18. T. Welton, *Coord. Chem. Rev.*, **2004**, *248*, 2459-2477.

19. P. Wasserscheid and W. Keim, *Angew. Chem. Int. Ed.*, **2000**, *39*, 3772-3789.

20. S. Pandey, *Anal. Chim. Acta*, **2006**, *556*, 38-45.

21. D. Wei and A. Ivaska, *Anal. Chim. Acta*, **2008**, *607*, 126-135.

22. M. C. Buzzeo, C. Hardacre and R. G. Compton, *Anal. Chem.*, **2004**, *76*, 4583-4588.

23. J. Berzelius, *Ann. Phys. Chem.*, **1826**, *82*, 369.

24. J. C. Marginac, *Ann. Chim. Phys.*, **1864**, *3*, 5.

25. M. v. Laue, *Physikalische Zeitschrift* **1913**, *14*, 1075.

26. W. L. Bragg, *Nature*, **1912**, *90*, 410.

27. J. F. Keggin, *Nature*, **1933**, *131*, 908.

28. M. Ammam, *J. Mater. Chem.*, **2013**, Advance Article

29. M. T. Pope and A. Müller, *Angew. Chem. Int. Ed.*, **1991**, *30*, 34-48.

30. A. Proust, R. Thouvenot and P. Gouzerh, *Chem. Commun.*, **2008**, *0*, 1837-1852.

31. J. Thiel, C. Ritchie, H. N. Miras, C. Streb, S. G. Mitchell, T. Boyd, M. N. Corella Ochoa, M. H. Rosnes, J. McIver, D.-L. Long and L. Cronin, *Angew. Chem. Int. Ed.*, **2010**, *49*, 6984-6988.

32. T. Yamase, *J. Mater. Chem.*, **2005**, *15*, 4773-4782.

33. B. Hasenknopf, *Front.Biosci*, **2005**, *10*, 275.

34. M. Aureliano, *Dalton Trans.*, **2009**, *0*, 9093-9100.

35. M. T. Pope, *Inorg. Chem. Concepts 8*, **1983**, Berlin: Springer-Verlag.

36. C. R. Sprangers, J. K. Marmon and D. C. Duncan, *Inorg. Chem.*, **2006**, *45*, 9628-9630.

37. J. J. Hastings and O. W. Howarth, *Dalton Trans.*, **1992**, 209-215.

38. J. Yan, D.-L. Long, E. F. Wilson and L. Cronin, *Angew. Chem. Int. Ed.*, **2009**, *48*, 4376-4380.

39. F. Taube, I. Andersson, S. Angus-Dunne, A. Bodor, I. Toth and L. Pettersson, *Dalton Trans.*, **2003**, 2512-2518.

40. C. Streb, *University of Glasgow*, PhD thesis, **2008**.

41. V. W. Day, M. F. Fredrich, W. G. Klemperer and W. Shum, *J. Am. Chem. Soc.*, **1977**, *99*, 6146-6148.

42. A. Müller and S. Roy, *Coord. Chem. Rev.*, **2003**, *245*, 153-166.

43. F. Zonnevijlle, C. M. Tourne and G. F. Tourne, *Inorg. Chem.*, **1983**, *22*, 1198-1202.

44. A. B. Bourlinos, K. Raman, R. Herrera, Q. Zhang, L. A. Archer and E. P. Giannelis, *J. Am. Chem. Soc.*, **2004**, *126*, 15358-15359.

45. S. Roy, *Comments Inorg. Chem.*, **2011**, *32*, 113-126.

46. A. Dolbecq, E. Dumas, C. R. Mayer and P. Mialane, *Chem. Rev.*, **2010**, *110*, 6009-6048.

47. Y.-F. Song and R. Tsunashima, *Chem. Soc. Rev.*, **2012**, *41*, 7384-7402.

48. Y. Leng, J. Wang, D. Zhu, X. Ren, H. Ge and L. Shen, *Angew. Chem. Int. Ed.*, **2009**, *48*, 168-171.

49. K. Li, L. Chen, H. Wang, W. Lin and Z. Yan, *Appl. Catal. A*, **2011**, *392*, 233-237.

50. H. Li, Z. Hou, Y. Qiao, B. Feng, Y. Hu, X. Wang and X. Zhao, *Catal. Commun.*, **2010**, *11*, 470-475.

51. Y. Qiao, Z. Hou, H. Li, Y. Hu, B. Feng, X. Wang, L. Hua and Q. Huang, *Green Chem.*, **2009**, *11*, 1955-1960.

52. Z. Sun, M. Cheng, H. Li, T. Shi, M. Yuan, X. Wang and Z. Jiang, *RSC Adv.*, **2012**, *2*, 9058-9065.

53. W. Huang, W. Zhu, H. Li, H. Shi, G. Zhu, H. Liu and G. Chen, *Industrial & Engineering Chemistry Research*, **2010**, *49*, 8998-9003.

54. A. G. W. Gianluca Bernardini, Chuan Zhao, Alan M. Bond, *PNAS*, **2012**, *109*, 11552-11557.

55. C. Zhao and A. M. Bond, *J. Am. Chem. Soc.*, **2009**, *131*, 4279-4287.

56. J. W. Sons, *Inorg. Syn.*, **1990**, 89-90.

57. S. Lin, W. Liu, Y. Li, Q. Wu, E. Wang and Z. Zhang, *Dalton Trans.*, **2010**, *39*, 1740-1744.

58. M.-H. Chiang, J. A. Dzielawa, M. L. Dietz and M. R. Antonio, *J. Electroanal. Chem.*, **2004**, *567*, 77-84.

59. F. Zonnevijlle, C. M. Tourne and G. F. Tourne, *Inorg. Chem.*, **1982**, *21*, 2742-2750.

60. K. F. Jahr, J. Fuchs and R. Oberhauser, *Chemische Berichte*, **1968**, *101*, 477-481.

61. K. F. J. J. Fu chs, *Z. Naturforsch. part B*, **1968**, *23*, 1380.

62. L. San Felices, P. Vitoria, J. M. Gutiérrez-Zorrilla, L. Lezama and S. Reinoso, *Inorg. Chem.*, **2006**, *45*, 7748-7757.

63. H. M. M. Hesse, B. Zeeh *Spektroskopische Methoden in der organischen Chemie*, *Thieme-Verlag*, *7. Auflage*, **2005**

64. G. M. Varga, E. Papaconstantinou and M. T. Pope, *Inorg. Chem.*, **1970**, *9*, 662-667.

65. Z. Wang, S. Gao, L. Xu, E. Shen and E. Wang, *Polyhedron*, **1996**, *15*, 1383-1388.

66. E. Riedel, C. Janiak, *Anorganische Chemie, de Gruyter-Verlag*, **2007**, *102*, 1371 and 1660.

67. P. G. Rickert, M. R. Antonio, M. A. Firestone, K.-A. Kubatko, T. Szreder, J. F. Wishart and M. L. Dietz, *J. Phys. Chem. B*, **2007**, *111*, 4685-4692.

68. K. A. K. Nageshewar, *Indian J. Chem. A*, **2010**, *45*, 635.

69. L. Dai, S. Yu, Y. Shan and M. He, *Eur. J. Inorg. Chem.*, **2004**, 237-241.

70. R. G. Larson, *The structure and rheology of complex fluids*, Oxford University Press, **1999**.

71. D. P. J. Dealy, *Rheology Bulletin*, **2009**, *78*, 16.

72. J. Hohenberger, K. Ray and K. Meyer, *Nat Commun*, **2012**, *3*, 720.

73. C. F. H. Mariusz Kozik, Louis C. W. Baker, *J. Am. Chem. Soc.*, **1986**, *108*, 2748-2749.

74. P. C.-R. Nieves Casai-Pastor, Geoffrey B. Jameson, Louis C. W. Baker, *J. Am. Chem. Soc.*, **1991**, *113*, 5658-5663.

75. M. K. Thomas L. Jorris, Nieves Casan-Pastor, Peter J. Domaille, Richard G. Finke, Warren K. Miller, Louis C. W. Baker, *J. Am. Chem. Soc.*, **1987**, *109*, 7402-7408.

76. *Lecture notes for "Applied Physical Inorganic Chemistry" Inorganic & General Chemistry lecture by Prof. Karsten Meyer in Summer 2008 at the Friedrich-Alexander university of Erlangen-Nürnberg*

77. G. F. T. Claude M. Tourne, *J. inorg. nucl. Chem.*, **1970**, *32*, 3875-3890.

78. K. Fukaya, A. Srifa, E. Isikawa and H. Naruke, *Journal of Molecular Structure*, **2010**, *979*, 221-226.

79. N. Haraguchi, Y. Okaue, T. Isobe and Y. Matsuda, *Inorganic Chemistry*, **1994**, *33*, 1015-1020.

80. R. D. Peacock and T. J. R. Weakley, *J. Chem. Soc. A*, **1971**, 1836-1839.

81. C. T. Rocchiccioli-Deltcheff, *R. J. Chem. Res. Synop.*, **1977**, *46*.

82. E. Riedel, C. Janiak, *Anorganische Chemie, de Gruyter-Verlag*, **2007**, *102*, 75.

83. B. L. Yujuan Niu, Ganglin Xue, Huaiming Hu, Feng Fu, Jiwu Wang, *Inorg. Chem. Commun.*, **2009**, *12*, 853-855.

84. A. Tézé, M. Michelon and G. Hervé, *Inorg. Chem.*, **1997**, *36*, 505-509.

85. Y. Niu, B. Liu, G. Xue, H. Hu, F. Fu and J. Wang, *Inorg. Chem. Commun.*, **2009**, *12*, 853-855.

86. F. Zonnevijlle, C. M. Tourne and G. F. Tourne, *Inorg. Chem.*, **1982**, *21*, 2751-2757.

87. P. G. Rickert, M. R. Antonio, M. A. Firestone, K.-A. Kubatko, T. Szreder, J. F. Wishart and M. L. Dietz, *Dalton Trans.*, **2007**, *0*, 529-531.

88. B. Dawson, *Acta.Crystallogr*, **1953**, *6*, 113.

89. H.-S. Liu, J. Peng and L.-X. Wang, *Zeitschrift für anorganische und allgemeine Chemie*, **2009**, *635*, 2688-2691.

90. R. Cao, K. P. O'Halloran, D. A. Hillesheim, K. I. Hardcastle and C. L. Hill, *CrystEngComm*, **2010**, *12*, 1518-1525.

91. K. Nomiya, Y. Togashi, Y. Kasahara, S. Aoki, H. Seki, M. Noguchi and S. Yoshida, *Inorg. Chem.*, **2011**, *50*, 9606-9619.